GIS in Real Estate

Geographic

Information

Systems

GIS in Real Estate:

Integrating, Analyzing, and Presenting Locational Information

Gilbert H. Castle III, Editor

Contributors

Anthony H. Hutchinson

John D. Landis

Bill Mundy, MAI

Michael L. Robbins

Stephen E. Roulac

L. Ried Schott, MAI

Grant Ian Thrall

Marcie B. Wallis

Bruce R. Weber, MAI

Appraisal Institute

Published in association with Adams Business Media/GIS World

APPRAISAL INSTITUTE®

Vice President, Educational Programs and Publications: Sean Hutchinson
Director, Content Development & Quality Assurance: Margo Wright
Manager, Book Development: Stephanie Shea-Joyce
Manager, Design/Production: Julie B. Beich
Production Associate: Michael P. Landis

For Educational Purposes Only

The material presented in this text has been reviewed by members of the Appraisal Institute, but the opinions and procedures set forth by the authors are not necessarily endorsed as the only methodology consistent with proper appraisal practice. While a great deal of care has been taken to provide accurate and current information, neither the Appraisal Institute nor its editors and staff assume responsibility for the accuracy of the data contained herein. Further, the general principles and conclusions presented in this text are subject to local, state, and federal laws and regulations, court cases, and any revisions of the same. This publication is sold for educational purposes with the understanding that the publisher is not engaged in rendering legal, accounting, or other professional service.

Nondiscrimination Policy

The Appraisal Institute advocates equal opportunity and nondiscrimination in the appraisal profession and conducts its activities in accordance with applicable federal, state, and local laws.

Printed in the United States of America

Library of Congress Cataloging-in-Publication Data

Castle, Gilbert H.

Geographic Information Systems : GIS in real estate : integrating, analyzing, and presenting locational information / by Gilbert H. Castle.

p. cm.

ISBN 0-92-215452-X

1. Real estate business—Data processing. 2. Real estate business—United States—Data processing. 3. Geographic information systems. I. Title II. GIS in real estate, integrating, analyzing, and presenting locational information III. Title: Integrating, analyzing, and presenting locational information

HD1380 .C37 1998

910'.285—dc21

98-47198

CIP

contents

Foreword xi

Preface xiii
by Gilbert H. Castle III
Purpose xiii
What Is a GIS? xv
Organization of the Text xvi
The Authors xvi

About the Authors xvii

Introduction 1
by Stephen E. Roulac

Chapter 1 5
An Overview of GIS Technology in Real Estate
by John Landis, PhD
How Does GIS Work? 5
GIS Functionality: What Can It Do? 8
Presentation and thematic mapping
Geocoding and address matching
Data queries and spatial queries
Database preparation and updating
Measuring distances and minimum paths
Buffering and point-in polygoning
Overlay and other polygon processes
GIS Uses in Real Estate 13
Appraisers
Brokers
Property developers
Investors and fund managers
Lenders
Market researchers and analysts
Shopping center owners and operators
Navigating the Ever-Changing GIS Marketplace 16
Street navigation: where am I?
Mapping add-ons
GIS viewers
Specialized map analysis tools
Full-function GIS
The newest frontier: maps and mapping over the Internet

Finding the Right Real Estate Data 19
Base maps
Real estate demand data
Real estate supply data
Transaction data
Regulatory and entitlement data
Image data
Organizing for a Competitive Advantage 21
The desktop model
The management information system model
The vertically integrated model
Conclusion 23

Chapter 2 24
Commercial Applications of GIS Technology
by Gilbert H. Castle III
Property Valuation 24
Applications and benefits
Case studies
Implementation considerations
The bottom line
Commercial Brokerage 29
Applications and benefits
Case studies
Implementation considerations
The bottom line
Institutional Real Estate 36
Applications and benefits
Case studies
Implementation considerations
The bottom line
Corporate Real Estate 42
Applications and benefits
Case studies
Implementation considerations
The bottom line
Additional Private Sector Applications 48
Data vendors
Mortgage underwriters
Developers
Consultants
Timber companies
Public Sector Applications 53
Asset management
Economic development
Government agencies selling GIS data
Conclusion 56

Chapter 3 57
Residential Real Estate and GIS
by Marcie Wallis and Anthony Hutchinson
The National Association of Realtors® 57
GIS Products for the Realtor® 58
Vendor-developed systems
Proprietary computer systems
GIS Technology Mishaps 60
HomeView Realty Centers, Inc.
Real Estate Information Network (RIN)
The Internet 61
The GIS Future for Residential Realtors® 63
Developing market territory
Locating potential home buyers
Locating homes for buyers
Optimal routing of "tours/caravans"
Listing presentations
Realtor® association applications
Conclusion 65

Chapter 4 66
Overview and Case Studies in GIS-Based Appraisal
by Michael L. Robbins, PhD, CRE
The Nature of the Real Estate Decision Process 67
What is real estate?
Feasibility analysis
Approaches to determining feasibility
Marketing—The Key to Development 70
Market research
Market data versus merchandising data
The ethics of fit and monopoly
The role of GIS
General Types of Spatial Data 75
Spatial measurement
Generation of buffers
Graphic area data
Attribute data to match with graphic area data
Point data
Attribute data to match with point data
Point to area data
Case Studies 78
Improved property analysis: Bear Valley area
Unimproved property analysis: the Morrison quad
Case study conclusions
Spatial Decision Support for Automated Residential Property Valuation 87
Purpose
Valuation methodology
Strategic objectives of automated valuation methodology
The working environment of automated valuation

Market demand
The production product
Life cycles
The role of GIS in automated valuation
Conclusion 95

Chapter 5 96
Application of GIS to Real Estate Appraisal Problems
by Bruce R. Weber, MAI
Chapter Overview 97
Use of GIS in Appraisal Review 98
Example 1. Dot density mapping: basic quantification of demand
Example 2. Calculation of areas and volumes: contaminated soils
Example 3. Use of point-in polygoning to find comparables
Example 4. Calculating overlapping areas: biologically valuable land
Example 5. Use of data queries: an industrial example
Use of GIS in Retail Analysis
Example 6. Large retail site
Combining GIS Databases
Example 7. Residential land example
Use of GIS for Business Cycle Analysis and Forecasting 116
The human model of forecasting
Example 8. The Phoenix office market
Real estate market forecasts
Example 9. Defining the relevant market area for a large apartment complex
How GIS Can Help Appraisers Respond to Criticism 121
Summary and Conclusion 122

Chapter 6 124
The Appraiser and GIS in a Litigation Context
by Bill Mundy, MAI, PhD, CRE
The Choice of an Expert 124
The Appraiser as an Expert 124
Inception
Data gathering and analysis
Summary 130

Chapter 7 131
Software, Data, and Management Considerations in Establishing a GIS/Mapping System
by L. Ried Schott, MAI, SRA
Similarities and Differences Between Mapping and GIS 131
Critical Success Factors 133
Software 134
Background
GIS categories
Mapping categories
Specialization
Summary and conclusions of GIS and mapping software

Data 145
Data sources and types
Data classifications
Aerials, photos, diagrams, and maps
Supplemental sources
In-house information
Final comments on data issues
Management Issues 153
Tangible returns
Intangible returns
Financial considerations
Vendor support
Staffing and assistance
Education and training
Implementation alternatives
Organizational Issues 161
Preimplementation studies
Answering questions
Final Comments 164

Chapter 8 165
Real Estate Information Technology: The Future Is Today
by Grant Ian Thrall, PhD
Residential Real Estate 166
Residential brokerage
Mortgage lending
Appraisal
Housing demand
Commercial Property 169
Office
Retail
Industrial
Implementation Factors 171
Geographic data
GIS software
Real estate information technology personnel
Conclusions 174

Appendix 1

Additional GIS Sources 175

Appendix 2

Color Maps 183

foreword

A geographic information system, which examines and displays data in a spatial context, is uniquely suited to the detailed analysis of real estate data and its persuasive presentation in reports. Although GIS has been around for a number of years, its use by real estate professionals is still limited. An adventurous few are ahead of the curve, using GIS systems to locate properties for home buyers, perform supply and demand analyses for retailers, and advise investors on the performance of their assets. Many other real estate practitioners are waiting for applications to expand and for hardware and software costs to come down before exploring the advantages of this exciting new tool.

As *GIS in Real Estate* will demonstrate, the wait is over and the time to act is here. With knowledgeable assistance and a reasonable financial investment, residential and commercial real estate brokers, appraisers, assessors, and investment analysts can tap into applications and insights that did not exist just a few years ago. The uses and benefits of GIS technology are so far-reaching, in fact, that they touch all the major players in the real estate industry—investors, developers, property owners and managers, service providers, market observers, and regulators.

Of course, the advantages of any new technology are meaningless unless you can put it to work for you. The difficulty is in the details, which is why this text takes the reader through all the decisions and problems encountered in setting up a GIS. It contains essential information on acquiring the proper hardware and software, making use of digital and attribute data, custom programming, and system integration. It even provides advice on staff training, the use of outside consultants, ongoing project management, and future GIS applications. Case studies are presented to show readers how their peers and competitors are making decisions—and money—by performing in-depth, sophisticated analyses from their desktop computers.

Joseph R. Stanfield, Jr., MAI, SRA
1998 President
Appraisal Institute

preface

by Gilbert H. Castle III

PURPOSE

Readers of this book will learn how Geographic Information Systems (GIS) can help real estate professionals to:

- Increase productivity, credibility, and corresponding profitability.
- Find out what the competition is doing.
- Stay in front of momentous changes in the real estate industry.
- Work cheaper, faster, and better.

Real estate is a game of information arbitrage. Since the economic depression of the real estate industry in the 1980s and 1990s, real estate professionals have learned to be increasingly data- and computer-intensive in their decision making. GIS is the computer technology that assembles, analyzes, and displays data based on property location. In short, GIS is superbly applicable to the information needs of the real estate industry.

The real estate industry is very large and very diverse. The total value of commercial and residential real estate in the United States in the mid-1990s has been estimated at $2.4 trillion and $8.7 trillion, respectively.[1] By comparison, the total value of corporate equities has been estimated at $4.1 trillion. Figure 0.1 illustrates the diversity of the industry, identifying six major categories and approximately 40 subcategories of real estate-related professionals.

Each of these industry players has unique GIS needs. The questions these individuals are asking include the following:

- Who else is using GIS technology in real estate? How long has it been used and how successfully?
- Can I operate a GIS myself, or do I have to hire a GIS technician?
- Where do I get the data? How much does the data cost? How often must it be updated?
- How easy is it to learn GIS? Can I retain my knowledge after several days or even weeks of not using it?
- Does a GIS pay for itself? How soon will I recapture my investment and in what ways?
- Can I buy off-the-shelf software and data that apply to my job, or do I have to customize a generic GIS?
- How does GIS differ from other computer technologies?

1. David J. Hartzell, Robert H. Pittman, and David H. Downs, "An Updated Look at the Size of the U.S. Real Estate Market Portfolio," *The Journal of Real Estate Research* (Spring 1994): 211.

Figure 0.1

Real Estate Industry Players

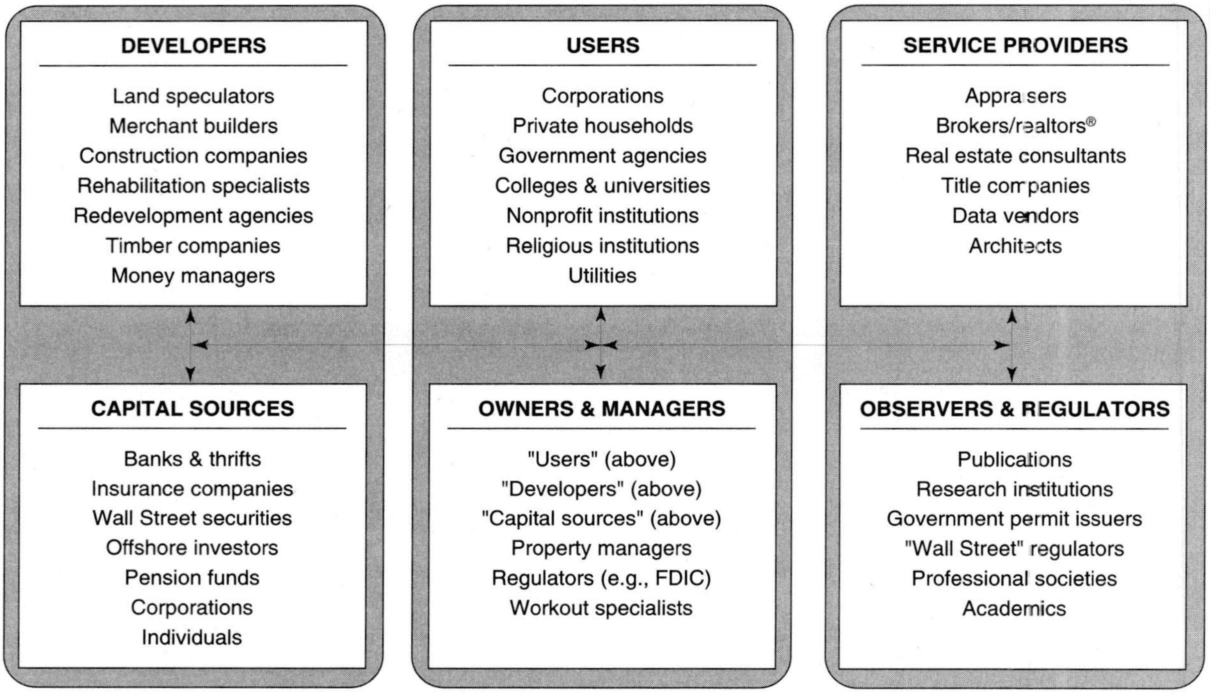

Source: Castle Consulting, San Francisco

- How many sources of GIS software, hardware, data, consulting services, customized programming, etc., are available? How do I find them and how do I choose among them?
- What is the likely risk-reward ratio? cost-benefit ratio? return on investment (ROI)?
- What are the best aspects of a GIS? What are the worst aspects?
- Can a GIS reside on a laptop computer? on a desktop computer? on a mini or mainframe computer? on a local area network, wide area network, and/or time-sharing system?
- How does a GIS relate to multimedia technology, including digital photographs, full-motion videos, and satellite imagery?
- Can I easily link a GIS to a desktop publishing system, e.g., to produce marketing materials? Can it be linked to my current management information system (MIS)? to spreadsheets such as Excel or Lotus 1-2-3? to non-GIS real estate software such as Pro-Ject, Skyline, and Dynalease?
- How is the Internet changing the delivery and use of GIS software and data?
- How do I get started?

These questions and others are addressed in this text.

Figure 0.2 shows the interfacing of the six major categories of real estate players their common analysis activities (brokerage, appraisal, etc.), and the products and services within GIS (software, hardware, data, etc.). The three-dimensional matrix consists of several hundred cells, each representing a "GIS in real estate" niche. Five years ago most of these matrix cells were empty; today most of the cells are active, but this exhortation is still valid: "Dominate a niche and get rich!" The bottom line is: Sooner or later you will be using a GIS, if you are not already. The only question is whether you will be a leader or a follower.

Figure 0.2

GIS Opportunities in Real Estate

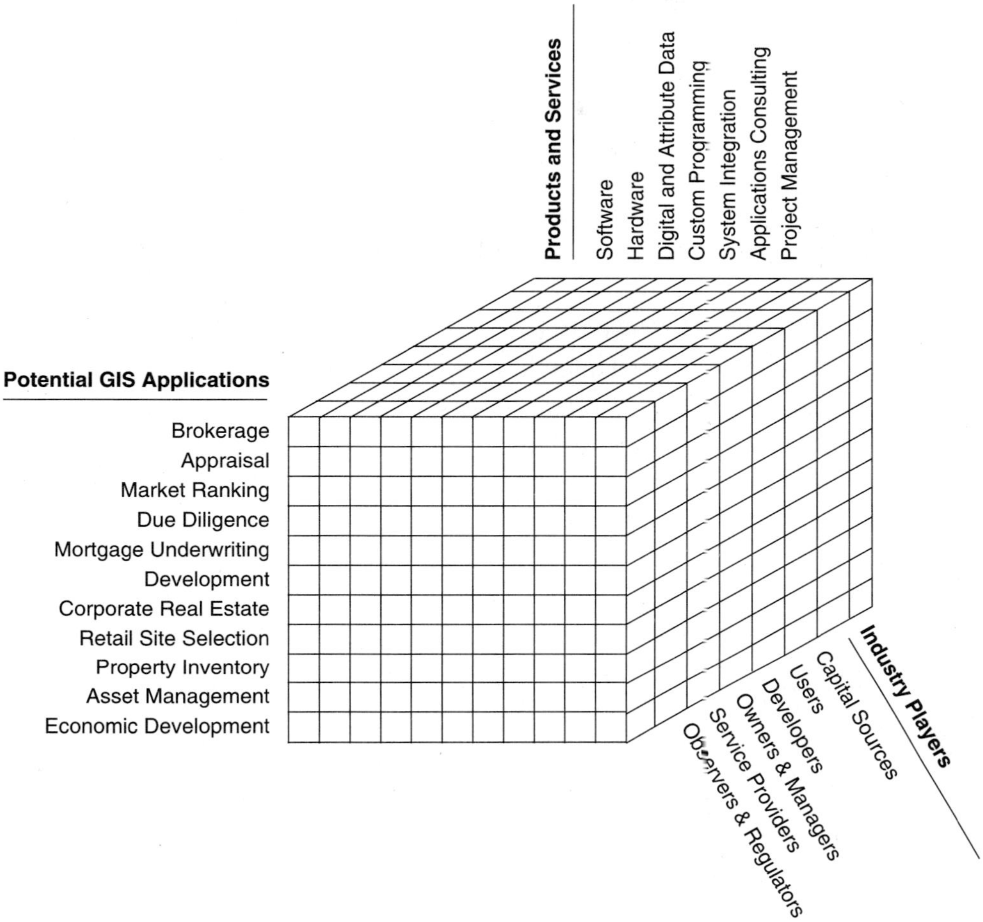

WHAT IS A GIS?

The first chapter in this text defines GIS. The remaining chapters flesh out this definition with case studies focusing primarily on valuation applications.

GIS encompasses distinct software, hardware, and data. Essentially GIS is a marriage between computerized mapping and database management systems. Stated differently, GIS offers valuable tools for location-based *visualization* and *integration* of innumerable types of data. Thus, an investment in GIS can be justified in terms of improved presentations and/or decision support.

GIS technology has been evolving for approximately three decades. GIS is now a billion dollar industry worldwide. The principal users continue to be government agencies with land administration responsibilities, e.g., the Bureau of Land Management at the federal level, natural resource departments at the state level, and assessor's offices at the local level. In the private sector, timber companies (the largest private landowners in the nation) were early adopters. A closely related technology—computer-aided design and drafting (CAD)—is extensively used by utility, telecommunications, and petroleum companies.

Although they still comprise a minority share of the market, business applications—in real estate, retailing, insurance, transportation, manufacturing, health care, etc.—are growing rapidly. The growth rate of GIS is remarkable given that the first-ever "GIS in Business" conference was held in Denver in March 1992. The first book on GIS, *Profiting from a Geographic Information System,* was published by GIS World, Inc. in March 1993.

GIS can be narrowly defined, but its impacts on the real estate industry are broad.

ORGANIZATION OF THE TEXT

GIS in Real Estate includes

- An introduction to GIS technology
- An overview of the technology and its applicability to commercial and residential real estate (Chapter 1)
- Case studies on how GIS is currently used in commercial and residential real estate, in both the private and public sectors (Chapters 2 and 3)
- The application of GIS technology to real estate appraisal (Chapters 4, 5, and 6)
- Acquiring, implementing and maintaining a GIS (Chapter 7)
- A look into the future of GIS in real estate (Chapter 8)
- Additional information on relevant Internet sites (Appendix 1)
- Color copies of some of the maps presented in the text (Appendix 2)

The book has value for both novices and experienced GIS users. Although no prior knowledge of GIS is assumed, readers who consider themselves "GIS literate" will find much of the information provided here useful.

THE AUTHORS

The authors who contributed to the text are all real estate professionals. Collectively, they have more than 100 years of experience in GIS. They are all fascinated by, and committed to, the ever-expanding use of GIS technology in real estate. While they have much in common, the contributing authors were selected for their diverse experiences and insights. Some are from academia, others work in the private sector, and two are active in trade associations. Several authors are involved principally with commercial real estate, while others have a residential focus. Theoretical versus practical, managerial versus technical, near-term versus long-term—these perspectives and a broad spectrum of other views are represented.

The authors have donated a significant amount of their valuable time to the creation of this book. They hope that its readers experience success in applying the knowledge they will gain from *GIS in Real Estate*.

about the authors

Gilbert H. Castle III, Editor, is the CEO of Castle Consulting in San Francisco, which focuses exclusively on the use of GIS in real estate. He has been a GIS consultant for 20 years. Prior to founding Castle Consulting, Mr. Castle was a regional director of the Roulac Group of Deloitte & Touche, the real estate consulting division of the big six accounting firm, where he was the principal resource nationwide for consulting tasks requiring expertise on GIS technology. Mr. Castle is a past president of the Urban and Regional Information Systems Association (URISA), North America's oldest and largest professional society for GIS practitioners. He has developed more than 150 publications and conference presentations, mainly on applied computer technology in land use decisions. Mr. Castle is the overseeing editor of *Profiting from a Geographic Information System,* the seminal text on GIS applications in the private sector, and he writes the real estate column for *Business Geographics* magazine. Mr. Castle has lectured on applied GIS technology at numerous universities, including the Massachusetts Institute of Technology, the University of Pennsylvania, Stanford University, and the University of California at Berkeley. Mr. Castle can be reached via e-mail at <gcastle@castleconsulting.com. His web address is www.castleconsulting.com.

Anthony H. Hutchinson was formerly a research analyst at the National Association of Realtors® (NAR) in Washington, D.C., in charge of GIS real estate application development for the membership. Mr. Hutchinson is now a research analyst with Fannie Mae.

John D. Landis, PhD, is an associate professor in the Department of City and Regional Planning at the University of California at Berkeley. He teaches GIS-related courses in planning methods, project development, public economics, and housing. He has authored many articles on the use of GIS in planning and land use and contributed to *Profiting from a Geographic Information System.* Professor Landis also teaches professional development courses for the Urban Land Institute (ULI) and is the principal author of *ULI Guide to Market Analysis,* which is to be published in 1999. John Landis' e-mail address is landis@ced.berkeley.edu.

Bill Mundy, PhD, MAI, CRE, is president and senior analyst at Mundy & Associates, a Seattle firm specializing in the valuation of conservation, preservation, and contaminated properties. He received his BA in agriculture from Washington State University and has an MA in urban economics and a PhD in market research and urban economics from the University of Washington. Mr. Mundy has published articles in various professional and academic journals. Bill Mundy's e-mail address is bill@mundyassoc.com and his web address is www.accessone.com/~mundy

Michael L. Robbins, PhD, CRE, has more than 25 years of experience teaching real estate appraisal at the University of Wisconsin at Madison and at Denver University. In addition to his academic activities, he has been a consultant to local and state governments, Alaska native corporations, private industry, and international agencies. Dr. Robbins is currently the CEO of GRAASroots Real Estate Counseling, the only company providing a GIS-based automated appraisal system to mortgage underwriters.

Stephen Roulac, chief executive officer of The Roulac Group, is a nationally recognized expert on strategic management, capital markets, securitization, institutional investing, and property analysis/valuation. He organized, led, and had major impacts on developing the national consulting practices of Kenneth Leventhal and Deloitte & Touche. As a leading academic, he has served on the faculties of Stanford Graduate School of Business, the UC Berkeley Architecture Department, and Hastings College of Law. Mr. Roulac is the author of more than 300 articles and a dozen books, many considered landmarks. Much in demand as a professional speaker, he has delivered presentations and speeches to some 400 organizations. Past president of the American Real Estate Society, Mr. Roulac holds a PhD in finance and strategy from Stanford; a JD from UC Berkeley; an MBA from Harvard; a BA from Pomona; and the CPA, CMC, and AICP designations. Stephen Roulac's e-mail address is sroulac@roulac.com and his Web page is at www.roulac.com.

L. Ried Schott, MAI, SRA is president of Schott Consulting Services in Los Angeles, a firm that provides real estate evaluations, GIS and mapping assistance, Internet consulting and Web site development, and other real estate services. With more than 20 years experience in various real estate fields, Mr. Schott has served as the research director of a pension fund advisor, director of consulting for a national real estate brokerage firm, vice president of an investment banking firm, and regional manager for an international appraisal firm. L. Ried Schott can be reached via e-mail at ried@schott.com or at his Web site address: www.schott.com

Grant Ian Thrall, PhD, is a professor of geography at the University of Florida in Gainesville. He has published more than a dozen books and over 100 articles on topics related to land markets, retail and industrial location, general spatial analysis, and geographic information technology. Dr. Thrall is also president and co-founder of Spatial Decisions & Analysis Inc., which designs and writes Microsoft Windows-based business GIS applications and advises business firms on the adoption and implementation of GIS technology. Dr. Thrall is editor of the real estate information technology section of the *Journal of Real Estate Literature,* published by the American Real Estate Society. He serves as editor and principal author of the "Shop Talk" and "First Impressions" columns in *Geo Info Systems* magazine. Grant Thrall's e-mail address is thrall@geog.ufl.edu. He can be reached at thrall@afn.org and his Internet URL is www.afn.org/~thrall.

Marcie B. Wallis is the managing director of marketing research at the National Association of Realtors® (NAR) in Washington, D.C. During her eight years at NAR, she has used GIS analysis extensively and has developed a GIS program for NAR. Now GIS is being used in many areas within the association as well as by Realtors® and Boards of Realtors® across the country.

Bruce R. Weber, MAI, has been a review appraiser for the Federal Deposit Insurance Corporation (FDIC) for many years and has special expertise in unusual and difficult-to-value properties. He has been asked to present innumerable papers at appraisal conferences around the world and was one of the first to publish articles on GIS in appraisal. Mr. Weber is also president of Market Foresight in Newport Beach, California. With more than 20 years of experience, he utilizes geographic information systems to quantify and forecast the demand for real estate and the financial risk resulting from soil contamination for consulting, appraisal, and appraisal review assignments. Bruce Weber can be reached via e-mail at bweber2@ix.netcom.com or at his Website www.marketforesight.com.

introduction

Stephen E. Roulac

The technology of geographic information systems (GIS) is integral to contemporary real estate practices and will become increasingly important in the twenty-first century. Like other emerging technologies, GIS promises to transform how business is done, challenge past practice, create new specialized practice, create specialized fields of professional work and make others obsolete, introduce new products and services into the marketplace, and threaten the competitiveness of those who do not adopt its power.

When I first realized that GIS would be an integral part of the real estate sector, in 1983, *TIME* magazine had just chosen the personal computer as "*TIME* Man of the Year." This honor recognized the extraordinary impact that the computer had had and would have on business and society, the form and content of work, and daily life. The *TIME* editors certainly were prescient, and now, a decade and a half later, even the most optimistic projection made in the 1980s seems hopelessly conservative.

The technology of GIS has also advanced over the past 15 years, but in a less obvious, less pronounced way. In business today, the question is not whether an organization and its workers use computers, but rather how, which hardware and software they rely on, and which applications they choose to employ. In contrast, GIS is in an *early adopter* stage. The critical strategic issues are *when* and *how* best to proceed to apply this new technology. Because the GIS advance has been less visible, and its present and prospective impacts have not been fully recognized, those who do embrace GIS have an opportunity to gain competitive advantages over those who do not.

This text provides valuable instruction and reference material for those who are unfamiliar with GIS, for those who are aware of GIS but have not yet implemented GIS technology, and for those who are already using GIS. It addresses the rationale behind GIS, presents examples of GIS applications, discusses the benefits of GIS and the daunting challenges to be faced in its implementation, and provides specific guidance on how to proceed. Readers will also find descriptions of hardware and software systems and the names of vendors and other resources to support GIS applications.

Fundamentally, GIS is a technology that presents and processes information in a spatial context, defined by the coordinates of location on the globe. GIS is both an analytical and communications tool, which allows people to understand and see relationships that would otherwise be difficult to grasp. With GIS, users can consider more information and see more complex relationships. GIS can display a prodigious quantity of information on the scale, concentration, frequency, and relationships between, and the attributes of, people, organizations, political jurisdictions, and communities. The relationships between the factors studied may be either expressly acknowledged or constructed purely for analytical purposes. GIS is integral to Internet applications, many real estate processes, and a myriad of processes that extend beyond real estate.

GIS has the ability to depict relational databases visually and to present disparate information linked by its connection to a common location. Before the development of relational database software, made available commercially by such companies as Oracle and Sybase, data were largely presented in a serial format. Relationships among the data were often ignored or discovered only by chance; when such relationships were addressed, the approach was often clumsy, lacking in both precision and comprehension. The advent of relational databases represents an extraordinary advance in quantitative marketing, allowing users to employ quantitative measurement of buying behavior and decision making in reaching operational, tactical, and strategic decisions.

Relational databases are at the core of GIS. GIS technology provides a processing mechanism to link data with location and then display the result visually. This visual display is important for purposes of both perception and communication. My esteemed colleague, the late James Graaskamp, advocated a visual approach to real estate, decrying the excessive reliance on text-exclusive sterile depictions of property, which is inherently rich in color and form. More so than any other technology, GIS communicates visually. Some of this emphasis on visual thinking is reflected in Chapter 4, which was written by Mike Robbins, who draws on his past experiences working with James Graaskamp on specialized appraisal applications.

GIS applications for real estate have become important for three reasons: the information explosion, technological innovation, and structural changes within the real estate industry. In the early 1980s, at the time I identified GIS as integral to the future of the real estate sector, the state of the art of real estate decision making was far from advanced and lagged behind most sectors of the U.S. economy. Four decades earlier, after the end of World War II, *Fortune* described the real estate sector of the U.S. economy as "the industry that the Industrial Revolution forgot."[1] Without the power of GIS technology, real estate decision makers are disadvantaged. They lack an essential tool to process the extensive information that must be considered to make informed decisions consistent with contemporary business practices and competing investment alternatives.

By the mid-1980s, the pace of growth and change in the real estate sector was racing ahead of the information infrastructure. The real estate sector was essentially without a managerial orientation. Many organizations had no appreciation for the roles of strategy, marketing (particularly consideration of competition and customer needs), information systems, organizational structures, and planning/control models. The building and picking, self-dealing, fee-maximizing, transaction-driven approach prevalent in the past was becoming increasingly out of tune with the institutional, fiduciary, and professional style of the dominant decision makers in the business today.

In an article co-authored with Gilbert Castle and Lloyd Lynford, I described the challenges to real estate decision making in an information era:

> Virtually every segment of society is more information-sensitive. Just as real estate has lagged in the application of management techniques and theories in the industrial eras, the real estate sector seems to be lagging in the application of information to real estate decisions in the information era. This is not to say that there has not been a proliferation of information vendors, involving data bases and analytic technologies, nor has there been a shortage of various organizations embracing information in their self-promotional activities. The trouble is that few participants appear to have the ability to discern and validate the information available, and much of what appears can only charitably be described as so much "noise" on the circuits, adding more to confusion than to clarity.[2]

While the topic of information explosion has received much attention, many real estate professionals have treated information applications with what might charitably be described as benign

1. "The Industry the Industrial Revolution Forgot," *Fortune* (1947)
2. Stephen E. Roulac, Lloyd Lynford, and Gilbert H. Castle III, "Real Estate Decision Making in an Information Era," *The Real Estate Finance Journal* (Summer 1990): 9

neglect. The processes they employed have reflected primitive technologies, lack of access to information, and little facility for processing the information they have somehow managed to acquire.

Too many real estate analysis reports of the 1970s and 1980s could be described as having as their unifying theme the binding that held different pages together. The majority of these reports were largely innocent of an integrating perspective; linkages were not made and implications were not drawn. Conclusions concerning rental rates, absorption, projected investment returns, values, and risk factors were essentially disconnected from the information presented on national economic forces, regional economics, highest and best use considerations, the supply of competing space, user demand, government regulations and incentives, and capital market forces. GIS, which emphasizes location-linked information and integrating technologies, holds great promise for overcoming the shortcomings and deficiencies of much of what passes for property analysis.

GIS is an important discipline for real estate from the perspective of technology applications. As the Internet and electronic commerce gain both share of mind and share of commerce, the systems by which they work are critical. GIS is a way to present representations of three-dimensional environments and to create virtual environments.

GIS is an extraordinarily powerful means of implementing alternative real-world, real-time applications. The advent of the Internet raises expectations for all activities performed outside the Internet. More simply, those who interact with the Internet take the experiences they have from Internet interactions and transfer these expectations to their non-Internet involvements. Thus, the stakes have been raised. All business people today must be more compelling, more persuasive, and more effective in how they do their work and how they present it to the public. GIS represents an integral technology which will help real estate professionals respond to these expectations.

Among the forces that make GIS a particularly appealing technology are the following:

- *Visual.* In a society that is increasingly visually oriented, a compelling presentation of information and complex relationships is essential. One benefit of GIS is its power to display significant content in a visually appealing and efficient format.
- *Time.* Many business people expect things to happen in very short, compressed periods of time. Society has sped up, and people are impatient if things do not happen right away. GIS can facilitate processing more information and implementing analyses more rapidly than can alternative, more conventional approaches.
- *Content.* With such a proliferation of information readily available through the Internet, people have access to more information and knowledge and their content expectations are higher. GIS can enrich the content of information presentations.
- *Entertainment.* With entertainment encroaching upon the traditional realms of communications, news, and retailing, businesses recognize that people increasingly want to enjoy, if not to be entertained by, their business dealings. GIS promises more entertainment and enjoyment than do traditional analytical approaches.

As important as GIS is, however, it is crucial to remember that it is a technology and is not, in itself, content. This distinction is significant because many who are first exposed to a GIS demonstration perceive that the information and data depicted are essentially of a *free good availability* status. Most fail to appreciate how challenging and resource-intensive it can be to organize information for use in a GIS application. At the same time, once the data investment has been made, the information can be used repeatedly and its application offers extraordinary competitive advantages.

The profound and pervasive structural changes shaping the real estate sector also contribute to the appeal of GIS. These structural changes are reflected in the following ways:

- Enterprises involved in real estate becoming ever-larger
- Ownership shifting from private to public hands
- Title being held in securitized form rather than directly
- Investment orientation shifting from a single property to large, multiple-property portfolios

These structural changes have led to the challenge to process more information concerning more assets and more relationships. This information is increasingly characterized by both more complexity and more interdependency and extends much more broadly in scale and outreach than such information has in the past. GIS is an integral means for dealing with structural change.

Some pundits talk about the demise of the tangible, the irrelevancy of place and space, and the primacy of the intangible. Indeed, not long ago the community where you live and your street address conveyed the primary message about your status. That is no longer true. Today your electronic-mail address conveys the prestige that once was attached to real estate. As Paul Saffo, director of the Institute for the Future, observes, "Your net address says volumes about who you are, about what community you hang in, and whether you are a cyber snob or a cyber hick." What is happening is the realignment of place, space, and time.

For all the growing importance and influence of the Internet and the intangible, a very real renaissance of place consciousness is underway. Individuals have more choices about where to live, work, shop, socialize, and be entertained, and the choices they make become increasingly important. More factors influence these choices. More considerations come into play and more disparate influences must be reconciled. This is where GIS excels insofar as it can distill, integrate, and display disparate factors that influence decision making in contemporary society.

With information availability exploding and information becoming increasingly important in business and society, and therefore ultimately in real estate, the real estate sector faces a profound dilemma. Real estate professionals are extraordinarily challenged to cope with the information they have access to. And, given the current state of affairs, how could they hope to cope with the future? The dilemma was captured well by the admonition from the Red Queen to Alice in Lewis Carroll's *Through the Looking Glass,* "You must run as fast as you can just to stay in the same place. If you wish to get ahead, you must run at least twice as fast." GIS can help those in real estate to run *twice* as fast.

Those in real estate who choose to ignore the rise of GIS, the fundamental benefits of its applications, and its powerful implications do so at their peril. *GIS in Real Estate* is a seminal text to guide those who would not just run fast enough to stay in the same place, but rather wish to run twice as fast to get ahead.

chapter 1

An Overview of GIS Technology in Real Estate

by John Landis, PhD

Real estate, the saying goes, is about "location, location, and location." Until recently, this statement was a simple truism. Although real estate professionals might talk about location, they could never really measure or manipulate it. GIS makes it possible not only to measure location but also to incorporate location into many different types of real estate decision making. For example:

- Appraisers can use GIS to determine how location affects property values.
- Commercial brokers can use GIS to profile and query their listings and to identify key sales, leasing, pricing, or occupancy patterns.
- Developers can use GIS to analyze metropolitan growth patterns, to prioritize potential development sites, or to conduct environmental and traffic studies.
- Institutional investors and real estate investment trust (REIT) managers can use GIS to track property performance between and within metropolitan areas.
- Lenders can use GIS to keep track of their existing loan commitments, to help underwrite new loans, and to meet the reporting requirements associated with the Community Reinvestment Act.
- Market researchers can use GIS to analyze demographic data, to determine realistic trade areas, and to examine the spatial pattern of product competition.
- Residential Realtors® can use GIS to compare property availability and home prices by neighborhood or area.
- Retail operators can use GIS to identify new store locations or to evaluate the performance of existing outlets.

HOW DOES GIS WORK?

In a nutshell, GIS is a system for encoding, retrieving, analyzing, and presenting geographic or spatial data. Spatial data are phenomena, observations, or events that can be coded to or by location.

Most GIS systems are designed around similar principles. They include one or more map layers tied to one or more databases. Map layers are made up of individual map features, which typically include points, lines (also known as *arcs*), and polygons (sometimes known as *areas* or *regions*). Points are single pairs of *x-y* coordinates.[1] Lines are connected pairs of *x-y* coordinates. Polygons are "closed" lines, which means that the first *x-y* coordinate is also the last.[2]

Each map feature is also an entry in an accompanying database,[3] and each entry can have multiple characteristics or attributes. For convenience, the database can be thought of as a spreadsheet, with the map features as spreadsheet rows and the feature attributes as spreadsheet columns. To illustrate these basic concepts, consider the following example.

Figure 1.1 is a GIS map of the United States encompassing the 48 contiguous states (polygons), interstate highways (lines), and the 50 largest cities (points). Each state appears in a GIS database, where its attributes can include its name, 1997 population, current population, and so on. Each interstate highway link also appears in the database, where the highway link's attributes can include its interstate designation (e.g., I-10), speed limit, total length, etc. Finally, each city appears in the database, where its attributes can include the city's population, employment base, annual municipal budget, size in square miles, and so forth. Also shown in Figure 1.1 are the principal longitude-latitude lines intersecting the 48 contiguous states, since longitude-latitude coordinates (and other earth-based coordinated systems) are the means by which geographic information systems organize and integrate all data.

Figure 1.1
GIS Map of United States

More advanced GIS programs also include something known as *topology*. Topology is a sort of map intelligence—i.e., individual map features "know" where they are relative to other features,

1. The *x-y* coordinates can be presented in absolute terms (e.g., {1,1}, {1,2}, {2,2}, {2,1}) or relative to some external reference system (e.g., latitude-longitude). Locating a GIS map according to an external reference system, usually on the basis of three points, is known as *registering*.
2. In most GIS packages, point features, line features, and polygon features are actually stored in different data files.
3. In some GIS packages, each type of map feature has its own accompanying database; in others, the attributes describing all three map feature types are located in a single database.

again based on longitude-latitude coordinates. Paper maps lack topology. In a paper map of the 50 states, California is next to Nevada only because it was initially drawn that way. In a GIS map of the 50 states, California and Nevada share a common series of *x-y* coordinates. Moreover, there is likely to be a column in the database that, for each state entry, lists the names of adjacent states. Thus one could actively query a GIS map for the names of those states adjacent to California. No such queries are possible with paper maps.

Many GIS users, typically those who want to use GIS for the purposes of map display, do not require topology. More advanced GIS users—those who want to query their maps, compare features between layers, or analyze market data—typically require some level of topology.

Figure 1.2 illustrates how the topology between multiple data layers and their respective attributes can be used to determine the zoning, floodplain status, traffic counts, and other characteristics of a specific property.

Figure 1.2

Spatial Processing by GIS

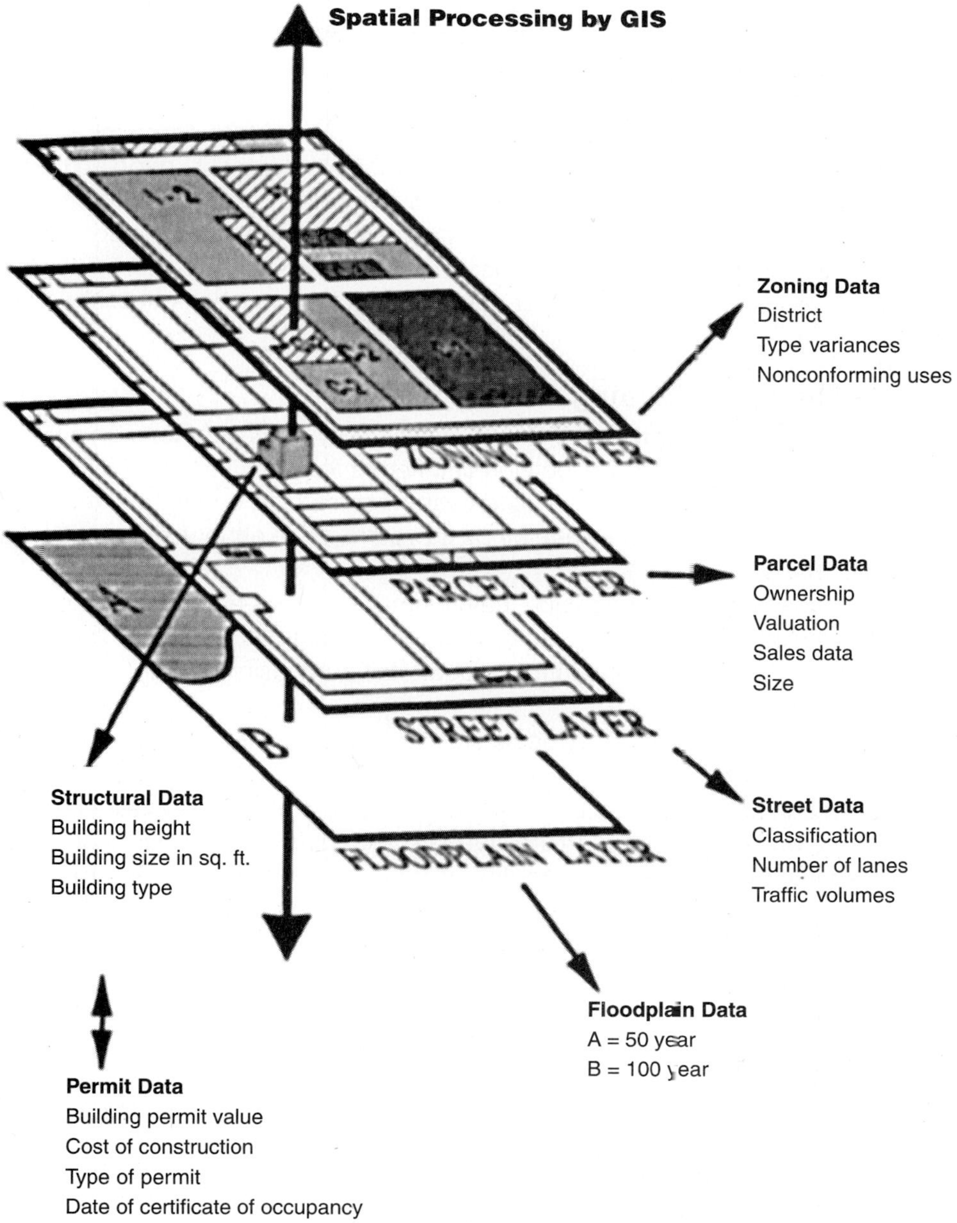

Source: *GIS Guidelines for Assessors,* URISA & IAAO, 1992

GIS FUNCTIONALITY: WHAT CAN IT DO?

Most GIS programs combine three basic types of capabilities. For perhaps 75% of all users, the most important GIS capability is *presentation mapping.* Presentation mapping includes the ability to show geographic features on a map, to assign data values and ranges to those features, to zoom in and zoom out to indicate various levels of detail, and to produce hard copy (paper) output. Just as spreadsheets and presentation graphics programs help viewers discern tabular or statistical patterns of information, presentation mapping helps viewers recognize spatial patterns of information. Presentation mapping is usually a one-way activity. The map functions as an output, a means of displaying information, not as an interactive query medium for accessing or organizing data.

A second level of GIS capability involves the use of maps as organizing tools for large databases. Mapping becomes a tool for users to query tabular databases, reorganize data into different spatial units, keep records of spatial data, and compare information across different features or databases. In a few cases—the dispatching of emergency services, for example—the map itself is the database.

A third level of GIS capability focuses on *spatial analysis.* Spatial analysis is the use of spatial data in logical or mathematical models for planning and decision-making purposes. Spatial analysis can take a variety of forms. For example, a real estate market analyst might use GIS to measure the distances between census tract locations and competing shopping centers. Once measured, the various distance estimates could be used in a retail gravity model to forecast shopping center market shares. Unlike presentation mapping and database organization, spatial analysis can rarely be systematized. It is a complex activity that requires an understanding of how the spatial relationships inherent in maps can be used with external theories and models for decision making.

The remainder of this section takes a closer look at each of the functions described and illustrates their application.

Presentation and Thematic Mapping

For most users, the biggest single attraction of GIS lies in its ability to present data on a map. Presentation maps can be used to show the location of different features and activities (e.g., the locations of recent home sales), or they can be used to show something about the attributes of those activities (e.g., the prices of recent home sales). Maps that summarize attribute information are referred to as *thematic maps.*

Polygon thematic maps consist of polygon features that are cross-hatched or colored to indicate the value of some attribute. A real estate market analyst might produce a polygon thematic map like the one shown in Figure 1.3 to illustrate average retail purchasing power (attribute) by census tract (polygon).

Point thematic maps consist of point symbols sized or colored to reflect the value of a particular attribute. Such maps have many real estate uses. For example, Figure 1.3 shows the locations of shopping centers in the San Francisco/San Bruno area. The centers are symbolized with mini-Parthenons of four sizes, which correspond to the standard shopping center classifications. The figure compares the spatial demand for retail facilities with the supply of retail facilities. The circle drawn on the map identifies a retail development opportunity, i.e., an area where significant retail demand exists, but there are currently no shopping centers. As it happens, the San Francisco 49ers plan to build a new football stadium and adjacent retail center at this location.

Point thematic maps can also be used to indicate occupancy rates in particular office buildings or the spatial patterns of home values. Figure 1.4 shows home prices in an area along the Connecticut shoreline. Such maps can also indicate where the customers of a particular community shopping center live and how much they spend. Lenders can use point thematic maps to summarize their loan portfolios, and institutional investors use them to keep track of quarterly property returns.

Figure 1.3
Density of Retail Purchase Potential

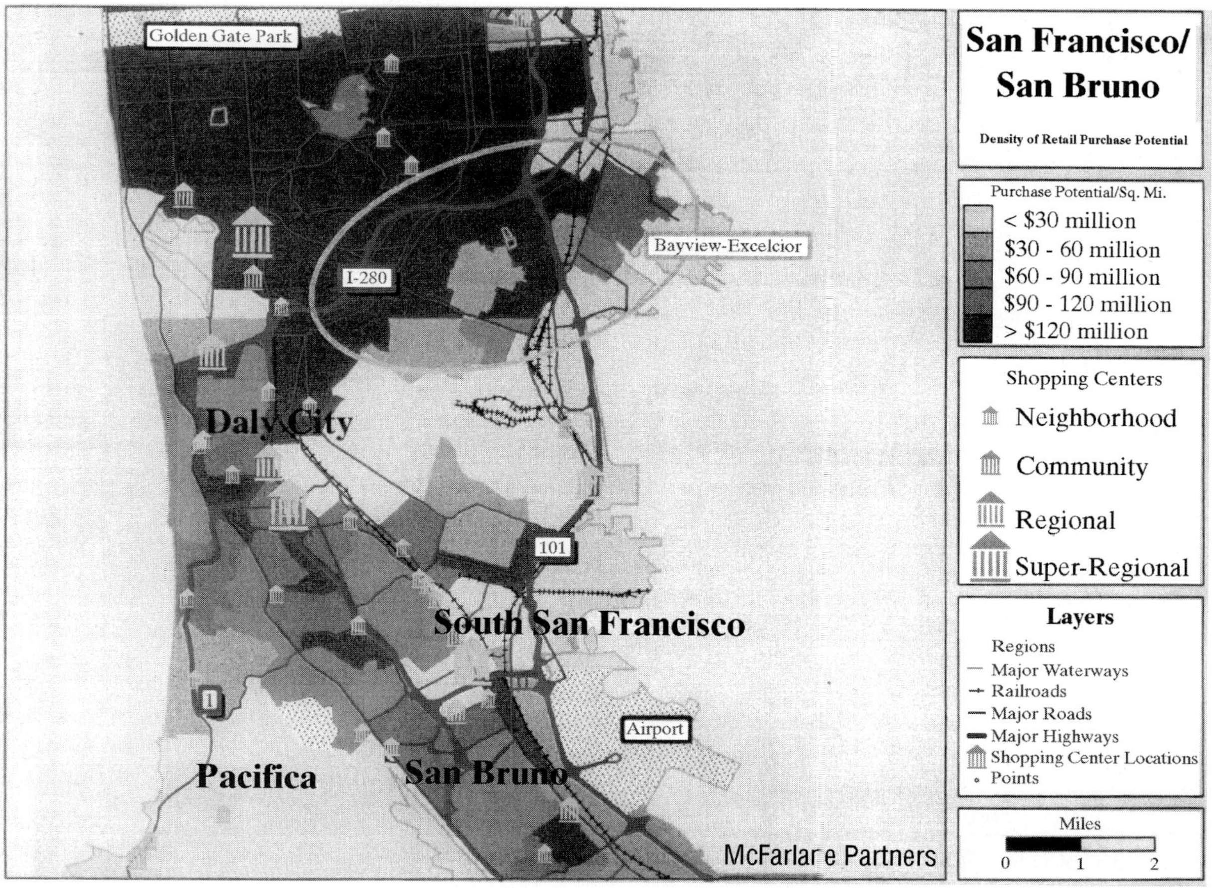

Line thematic maps consist of line segments of varying widths or colors which represent an attribute value. Line thematic maps are widely used for highway analysis (e.g., to show roadway congestion levels) and for environmental analysis (e.g., to show stream flows). They are less commonly used in real estate.

Geocoding and Address Matching

Geocoding is a process in which features are automatically, rather than manually, located on a map. Geocoding typically takes two forms: address matching and centroid location.

Most properties have addresses, so they can usually be address matched and depicted as points on a background street map. By comparing property street numbers to the street number ranges associated with particular street links, GIS programs can automatically position properties at or near their actual locations.[4] Address matching was used to prepare Figure 1.4.

Instead of street addresses, some properties are referenced according to their zip code zones or census tracts. Most GIS programs can geocode properties to the centroids of their respective zip code zones or census tracts. Once properties have been address matched or geocoded, they can be easily queried or thematically mapped.

Data Queries and Spatial Queries

Consider a commercial real estate broker who wants to prepare two separate thematic maps of asking rents for Class A and Class B office properties. The broker starts by using GIS to address

4. The accuracy of computer-based address matching depends on how well the street network is coded.

Figure 1.4
Thematic Point Coding of Sale Price

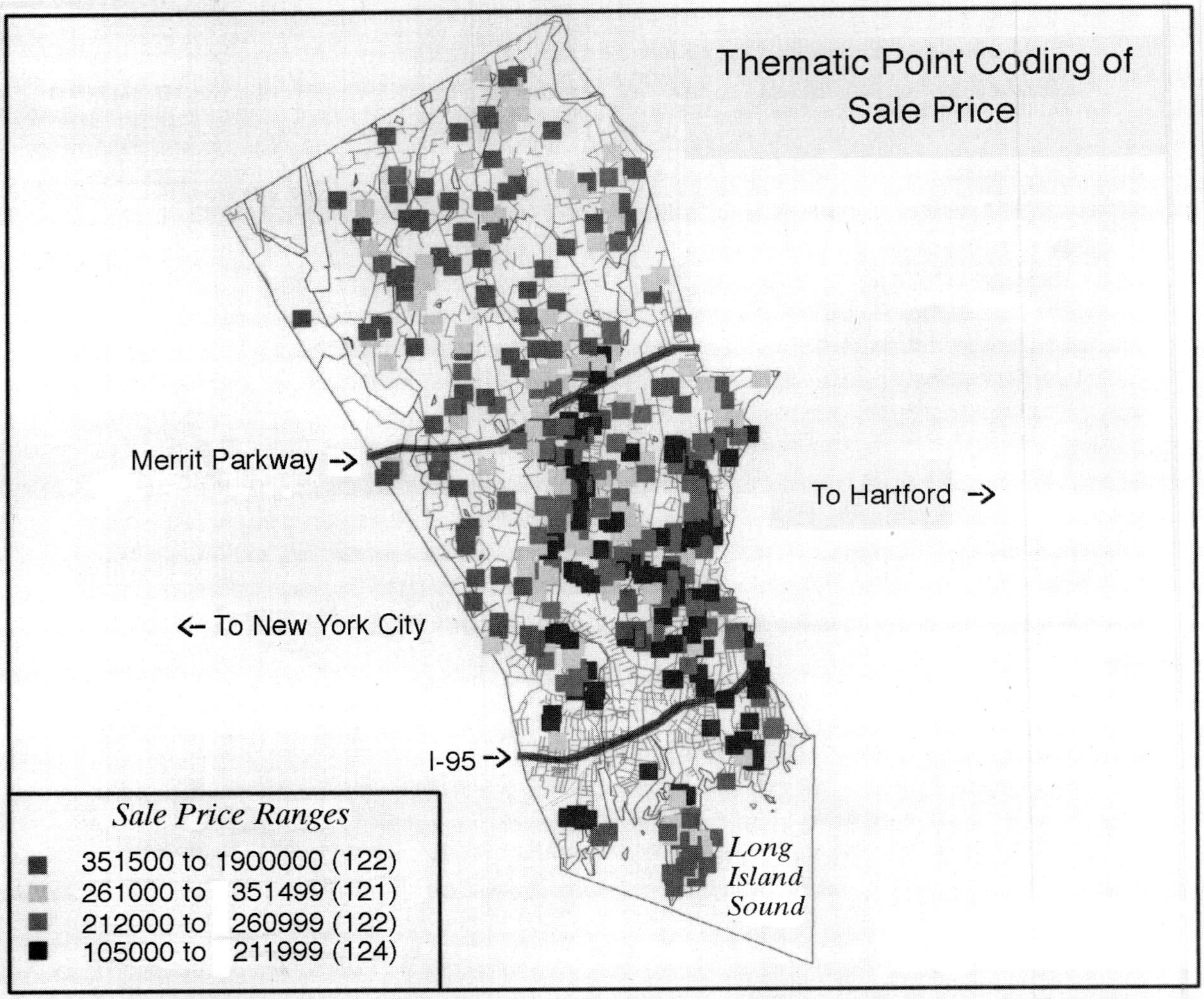

match a list of all office properties, regardless of class, to a street map. Next, the broker queries the database to display the locations and rents of Class A office buildings only. Once this first map has been printed, the broker separately queries the database to display the locations and rents of the Class B office buildings. We call this operation data querying or "show-me" querying. The broker is using GIS to "show" the results of a database query. Most GIS programs are able to undertake complicated, multiple-attribute data queries such as, "Show me the locations of only those Class A office buildings that are less than 10 years old, have available free parking, and have asking rents of less than $1.50 per square foot."

In a data query, the user queries the database and presents the query results in map form. In a *spatial query,* the user employs the map as a query tool to obtain a list of map elements that meet particular spatial criteria. Suppose that the commercial real estate broker discussed above wants to identify only those office properties that are within a quarter-mile of a particular downtown intersection. As a first step, the broker displays all the office buildings in the database. Next, using a mouse or pointing device, the broker draws a circle with a quarter-mile radius around the intersection of interest. The office buildings within the quarter-mile circle are automatically highlighted, and a list of building owners and addresses can quickly be printed out. Figure 1.5 depicts the process as applied by the brokerage firm of Pizzuti Development, Inc. with their GIS and property database for Columbus, Ohio.

Figure 1.5
Spatial Query of Office Properties

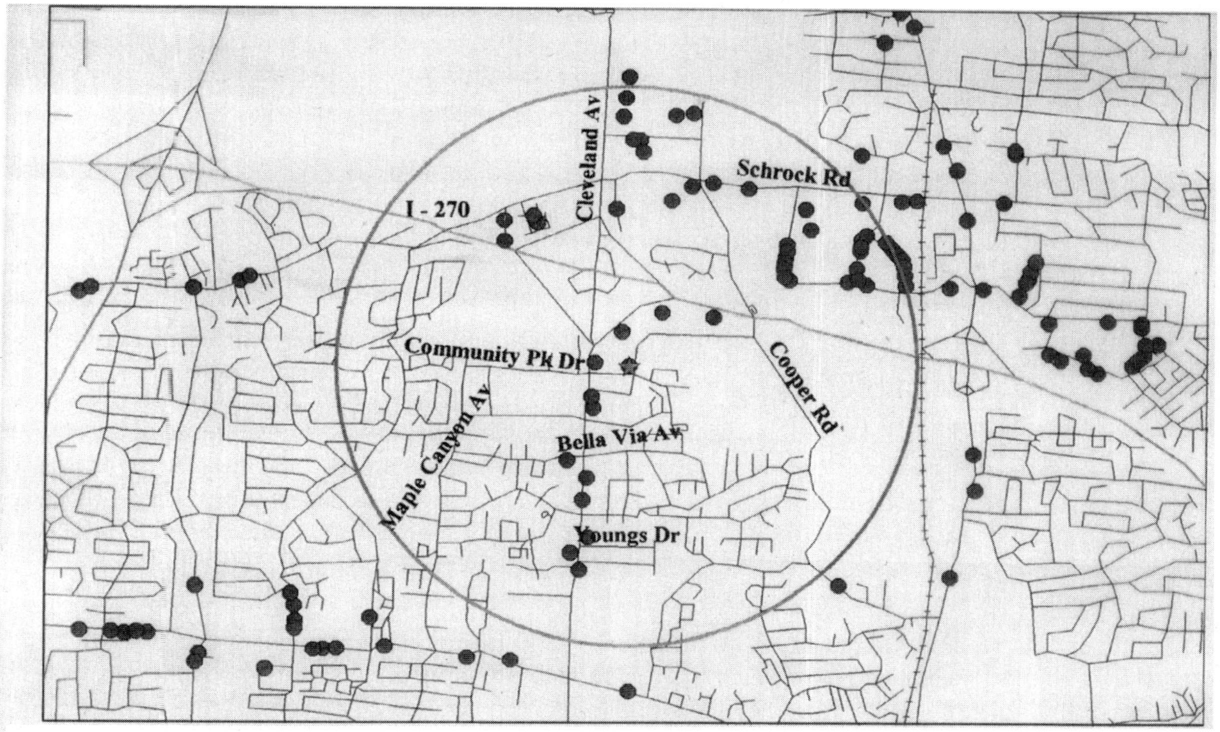

PIZZUTI

Office Building Information

Building No. 176

Park Name
Corporate Exchange Office Park
Building Name
Corporate Exchange IV
Address
2400 Corporate Exchange Dr.

City	State	Zip
Columbus	OH	43231

Gross Sq. Ft.	Available	Contiguous	Loss Factor	Smallest Div.
90,891	4,700	3,607	10.7%	1,850

Rent	Oper. Exp.	Terms	Vacancy
$13.50 To $13.50	6.05	Net	5%

Tax Dist.	Parcel	
600	183730	7/12/94

Project Status	Type Construction	Year Buil	No. Stories	Tenancy	Office Type	Ask Price
Existing	Glass	1986	3	Multi	Office	$0

Owner Information
Corporate Exchange Bldgs
Address Phone
383 S Third St.

City	State	Zip
Columbus	OH	43215

Sprinklers	☐	Quadrant	1
Year Reno.	0	Code	A
Transfer #	4	Taxes:	$138,112
		psf:	$1.52

Binns #	Grantor	Grantee	Sale Date	Amt. Sold	Per Sq. Ft.	Lot Size
0	2400 Realty Corp	Corporate Exchange	11/5/93	$14,500,00	$159.53	5.02
0		2400 Realty Corp	9/22/89	$	$0.00	5.02
0				$	$0.00	0

Spatial queries are also called "tell-me" queries because the map is used to identify, or "tell" about, those database records that meet a particular spatial criteria. In addition to circles, most GIS programs can identify map features within a rectangle or irregularly shaped, user-specified polygon.

Database Preparation and Updating

Real estate information is inherently dynamic. All of the data that real estate people use and rely on (e.g., home prices, commercial rents, total retail outlet sales, job and population growth rates, purchasing power figures, absorption rates) are constantly changing. The most up-to-date, spatially specific information is needed to make the right real estate decisions.

GIS can be especially helpful in preparing and updating real estate databases. Automatic geocoding procedures can facilitate the updating of property lists. Finding a feature on a map, pointing to it with a mouse, and then entering new or updated information is often far easier than searching and updating a tabular database.

Just as feature attributes such as price, rent, and occupancy change over time, so do the features themselves. Parcel boundaries may change. New stores or subdivisions may be developed. A REIT may acquire new properties in a different metropolitan area. GIS provides a convenient framework for entering and updating both the spatial and attribute characteristics of real estate data.

Measuring Distances and Minimum Paths

Distance analysis is the ability to calculate distance from a map, either aerially ("as-the-crow-flies") or along a transportation network. The ability to measure map distance accurately is essential in the determination of real estate market areas.

"What's the fastest way to get there from here?" This question is commonly asked in both social and business situations. To answer it one needs to go beyond the simple measurement of distance to find the shortest route, or minimum path, along a transportation network. Minimum path procedures are commonly used to identify competing retail trade areas.

Consider a case in which three community shopping centers compete for customers from 25 census tracts. Assume that shoppers are willing to travel no further than three miles from their homes to the nearest community shopping center. GIS can be used to identify quickly which census tracts fall within three miles of each community shopping center, measured in terms of minimum roadway distance. Although this type of analysis could be undertaken without GIS, it would be extraordinarily cumbersome and time-consuming. Minimum-path measurements are also required when trade areas are identified using a gravity model, a more realistic technique that considers spatial competition.

Nearly all GIS packages can measure point-to-point distance one at a time, some can estimate multiple distances or network distance, and a few can identify minimum network paths.

Buffering and Point-in Polygoning

A *buffer* is a zone of fixed width drawn around any map feature. A buffer around a point takes the shape of a circle, while a buffer around a line takes the form of a corridor. A buffer around a polygon is a larger polygon of similar shape.

Buffering is essential to establish *zones of influence.* A shopping center developer, for example, might want to count the number of households within a two-mile radius of a proposed neighborhood shopping center. Similarly, a real estate broker might want to buffer a recent home sale and send letters to the owners of neighboring properties to solicit new business.

The ability to buffer complex geographic features varies widely among GIS packages. While most packages can buffer points, lines, and simple polygons, only a few programs are capable of accurately buffering systems of concave (inwardly curving) polygons.

For example, assume that a lending institution is interested in using GIS to determine the total size and average value of outstanding mortgage debt by census tract. If the mortgages can be listed by address, the lender will first address match each loan to its street location. Next, using a GIS procedure known as *point-in polygoning,* the lender automatically assigns each mortgage (point) to a census tract (polygon). Finally, the lender computes the sum and average mortgage value associated with each tract.

As another example, consider a pension fund that is interested in comparing its real estate portfolio to known earthquake hazards. Again, the first step is to address match the properties to their street locations. Next, the fund managers would buffer known earthquake fault lines; each fault buffer would take the form of a polygon. By determining which properties (points) fell inside which fault buffers (polygons), the pension fund could determine its total exposure.

Point-in polygoning, like address matching, is a GIS procedure commonly used in many types of real estate analyses. Most GIS viewers and all full-function GIS have point-in polygoning capabilities.

Overlay and Other Polygon Processes

Overlay refers to the ability to compare map features and attributes spatially across two or more spatially coterminous map layers. Overlay procedures can take a number of forms:

- Intersection, in which only those areas and attributes common to both layers are identified;
- Union, in which all of the areas (and their attributes) included in both layers are identified;
- Clip, or extract, in which the common features shared by two or more layers are extracted;
- Erase, in which features not shared by two or more layers are erased; and
- Calculation, in which the attribute values of two or more polygon layers are used to compute new attribute values.

The intersection, union, clip, and erase functions typically involve making changes in feature geometry; calculation does not.

Overlay is the operation most people think of when they think of GIS. Among real estate applications, overlay is most frequently used in appraisal. Consider the case of an appraiser who is trying to determine the loss in site value associated with highway noise. As a first step, the appraiser buffers the highways to identify an appropriate noise corridor. This corridor, which is a polygon, can then be superimposed on a GIS-based parcel map. The geometric intersection of the noise corridor with the various parcels indicates the area of reduced value.

All GIS programs can visually superimpose one map layer over another. Many GIS programs can undertake simple overlays—i.e., show two or more layers that have exactly the same polygon geometry. Only the most powerful GIS packages can overlay layers that have different polygon geometry—e.g., superimposing floodplains onto a zoning map.

GIS USES IN REAL ESTATE

At first glance, many of the GIS functions described above would seem more relevant to university researchers or government agencies than to real estate professionals. However, there are many real estate activities, functions, and decisions that can be facilitated by the appropriate use of GIS (see Table 1.1).

Table 1.1

Summary of GIS Uses by Real Estate Activity

Key: A-M: address matching; DU: database updating; DQ: data query; SQ: spatial query; Buf: buffering; Dist: distance calculation; PiP: Point-in-polygoning; TM: thematic mapping; GC: geocoding

Real Estate Activity	GIS Use	GIS Functionality	Map Feature & Data Type
Appraisers	• Maintain database of comparable properties	A-M, DU	Property listings (points)
	• Identify nearby comparable properties	DQ, SQ, Buf, PiP	Property listings (points)
	• Measure and estimate value of proximity to amenities and public facilities (e.g., freeway interchanges, parks)	Dist, Buf, statistical modeling	Public facilities (points, lines, polygons)
Brokers	• Summarize available properties in map form	TM	Property listings (points)
	• Query map for specific property characteristics	DQ, SQ	Property listings (points)
	• Identify typical and unusual properties with respect to rent levels, prices, occupancy, or features	DQ, SQ, Buf, PiP	Property listings (points)
Developers	• Identify growth areas	TM	Census tracts, zipcodes (polygons)
	• Identify environmental and entitlement constraints to development	TM, Dist, Buf, PO	Parcels, environmental features (points, lines, (polygons)
	• Prepare environmental documents	TM, Dist, Buf, PO	Parcels, environmental features
	• Site and infrastructure planning	GC, TM	Parcels, infrastructure
Investors & fund managers	• Maintain map database of property holdings	DU, A-M, GC	Property listings (points)
	• Graphically compare property performance	TM, DQ, SQ	Property listings (points)
	• Identify areas for property acquisition/sale	TM	Cities, census tracts (points, polygons)
Lenders	• Maintain map database of loans	TM, A-M, DU	Loan listings (points)
	• Analyze geographic patterns of loan activity	DQ, SQ, Dist, Buf, PiP	Loan listings (points)
	• Improve loan underwriting	TM, DU, PiP	Loan listings (points)
	• Identify branch market areas	Dist, Buf	Loan listings (points)
Market researchers & analysts	• Identify and analyze population, job, and income trends by area	TM	Places, census tracts, zipcodes (polygons)
	• Identify project market areas	Dist, Buf, PiP	Census tracts, zipcodes (polygons)
	• Identify and evaluate competitive properties	A-M, GC, DQ, SQ	Property locations (points)
	• Determine contribution of location to property price or rent	Dist, Buf	Property locations (points)
Sales	• Keep track of prospects	TM, DQ, SQ, DU	Prospect addresses (points)
Shopping center owners & operators	• Keep track of store performance	TM, DQ, SQ, DU	
	• Identify market areas and potential locations for new stores	Dist, Buf, PiP	

Appraisers

Appraisal involves estimating market value using multiple comparable properties. GIS can improve appraisal in two ways. First, because GIS couples map and database functions, it can help appraisers keep track of hundreds or thousands of different properties and quickly determine which properties are locationally comparable. Second, when coupled with statistical procedures such as multiple regression, GIS can help appraisers estimate the additional property value associated with proximity to neighborhood amenities such as parks, transportation facilities (i.e., highway interchanges or mass transit stations), and nearby shopping opportunities as well as negative features such as dumps. In a recent study of home values in the San Francisco Bay Area, GIS was used to measure street distances from 1,200 single-family homes to the nearest freeway interchange and mass transit station. These measurements were added to a statistical model that compared residential sale prices by home size, age, and neighborhood quality. Home prices were found to be positively influenced by proximity to a transit station (i.e., home prices increased $2.00 per meter of proximity to a transit station) and negatively influenced by proximity to a freeway interchange.

Brokers

Brokers representing buyers and tenants need to be able to match the specific space and locational requirements of their clients to the properties available in the market. Brokers representing sellers and building owners need to be able to differentiate the properties they represent from the rest of the market. GIS can facilitate both of these activities.

Current property data are increasingly available via modem or over the Internet. Property listings can be downloaded, address matched to a street network, and then mapped. GIS can be used to query the characteristics of individual properties rapidly (e.g., "How many square feet of office space are currently available in the building I am indicating on the map?") or to indicate on a street map the locations of those properties that meet a client's specific needs (e.g., "Which office buildings have at least 5,000 square feet of space currently available?"). In addition to retrieving numerical information regarding rents, prices, square footage, and amenities, an increasing number of GIS packages can retrieve image information, including floor plans and views of building exteriors.

Property Developers

Developers make money by buying properties that are undervalued, developing them at a reasonable cost, and selling them for prices at or above market levels. Most developers find such properties on an individual basis, but GIS can improve the scope and efficiency of the search process. GIS-based thematic maps can help developers better understand where and why population and job growth is occurring. If information on recent land sales is available, GIS can help identify areas where land is underpriced or overpriced. GIS maps of soil quality, slope, wetlands and floodplains, and local zoning can help developers form a more accurate picture of the costs of developing a particular property. The same maps can be used to prepare the required environmental documentation and permit applications. Finally, developers can use GIS to track property sale prices.

Investors and Fund Managers

Investors and fund managers need to be able to summarize the financial performance of their portfolios quickly, and thematic maps provide an easy means to do this. GIS can also help investors and fund managers identify any geographic reasons why specific properties may be performing better or worse than expected. For example, an apartment property may be suffering from falling cash flows because it is located in a declining neighborhood. Alternatively, a decline in area vacancy rates may cause another property's net operating income to rise. GIS maps allow a portfolio manager to identify and track these income relationships from a distance.

Lenders

Real estate lenders usually have a good idea of the performance of their loan portfolios by metropolitan area or city, but they may not know how their loan portfolios vary by neighborhood or census tract. GIS makes it easy for lenders to take a local view. They can:

- Summarize their loan portfolios at the census tract or even census block level;
- Identify whether their nonperforming loans follow a geographic pattern;
- Compare basic lending criteria (e.g., average loan-to-value ratios, debt-coverage ratios, income-to-mortgage debt ratios); and
- Identify areas where lending risk has been unfairly overestimated (red-lining) as well as areas where risks have been systematically underestimated.

In short, GIS can be an essential tool in loan underwriting, in loan monitoring, and in demonstrating compliance with federal community reinvestment requirements.

GIS can also contribute to the retail side of lending. GIS-based site selection models can help lenders identify where they have too many branches and where they have too few. Similarly, GIS can be used to profile neighborhood demographics and help lenders determine which loan products should be offered at which locations.

Market Researchers and Analysts

GIS has more potential uses in market research than in any other real estate activity. Thematic maps can be used by real estate market analysts to identify, analyze, and present fundamental demographic, employment, and income trends.

GIS makes traditional, circle-based estimates of retail and office trade areas obsolete. Using the distance and routing capabilities of GIS, market analysts can now accurately estimate how many potential retail customers live within a certain distance or travel time of a retail center. Even better, market analysts can couple GIS-based measurements of distance and travel time with sophisticated retail market share models to estimate rates of retail market capture and/or leakage more accurately. Similarly, GIS can help market analysts estimate the likely effects of competing projects on market share, absorption rates, or competitive rents before project development or construction begins. Using statistical models similar to those used by appraisers, market analysts can use GIS maps to help residential developers establish pricing schedules and anticipate how neighborhood amenities and features will affect property prices.

Shopping Center Owners and Operators

Shopping center owners and developers typically have to master two types of businesses: the business of shopping centers and the business of retailing. In other words, a shopping center will only succeed if its retail tenants can attract sufficient customers willing to spend enough money. GIS can help shopping center operators and retailers identify sites that are underserved or overserved relative to the demand for different types of retail goods. With GIS, retailers can more easily identify potential retail locations, and shopping center operators can identify appropriate retailers for particular locations. On a more basic level, GIS-based thematic maps can be used by shopping center operators and retailers to monitor sales and income performance and to identify which outlets are outperforming or underperforming their demographic sales base.

NAVIGATING THE EVER-CHANGING GIS MARKETPLACE

Pushed by more powerful hardware and expanding user needs, the computer mapping marketplace continues to evolve. The current marketplace consists of five segments, each of which will be described below:

1. Simple street navigation
2. Mapping add-ons
3. Viewers
4. Map analysis tools
5. Full-function GIS

Moreover, an ever-increasing amount of GIS functionality and data are available over the Internet and the World Wide Web.

Street Navigation: Where Am I?

Sometimes we care only about location. Lost business travelers or brokers planning a tour of prospective properties care much more about finding particular locations than understanding or analyzing them. What these users want out of GIS is navigational assistance. Such assistance is now being provided in many convenient formats at ever-lower prices. Microsoft, for example, sells a single AutoMap CD for $49 that includes navigation software and navigable street maps of the 100 largest cities in the United States. Similar products are available at comparable prices from Rand McNally, DeLorme, Thomas Brothers, and others.

Navigation programs do not include convenient links to attribute data, so they cannot really be classified as GIS. Nevertheless, they increase our use of and comfort with maps, map-based data, and spatial relationships and thus provide a convenient gateway into the richer world of GIS.

Mapping Add-ons

Mapping add-ons are one level up from single-function navigational tools. They include thematic mapping procedures together with maps and data that work with popular spreadsheet programs. Both Lotus 1-2-3 v.5, and Excel for Windows 95, for example, include mapping add-ons for creating thematic maps of U.S. states and counties. Other mapping add-on programs currently include MACON USA and Maplinx.

Mapping add-on programs can be used to create terrific multicolor thematic maps at any scale and to identify specific map features and associated data. However, because they lack even a rudimentary form of topology, mapping add-ons typically cannot process compound spatial or data queries, compare features across layers, or be used for spatial analysis. Nor can they be used to edit maps or add new map features.

Who should consider mapping add-ons? Experienced spreadsheet users looking to map their data by state, county, zip code, census tract, or street location will find add-ons useful as will individuals who want to prepare their own presentation maps but do not have the time to learn an entirely new package.

GIS Viewers

One step up from mapping add-ons are GIS viewers such as ArcView, AtlasGIS, MapInfo, and Maptitude. The initial versions of these products (circa 1993) were designed to help users quickly construct multilayer thematic maps, address match point data, and undertake complicated map and data queries, including "near" and "buffer" queries. More recent versions have added rudimentary map-editing features, the ability to import photographic (or raster) images as background layers, and even some spatial analysis tools. More advanced GIS viewers allow users to edit underlying map features, conveniently store and update maps with thousands of features and data attributes, and aggregate existing map features and data.

Most GIS viewers are designed for use with existing off-the-shelf maps or maps digitized by third-party vendors. They are not intended for users wanting to generate their own maps.

Designed around the capabilities of desktop computers, the GIS viewer segment has become the largest and most active part of the GIS and computer mapping market. Because of their popularity, GIS viewers emphasize ease-of-use and broad functionality with well-designed, flexible window-based interfaces that cross different computing platforms. They work with a wide variety of color and black-and-white printers. A broad array of U.S. and international maps and attribute data sets are available off-the-shelf at reasonable prices. GIS viewers can be easily customized using macros or scripts, and because file translation programs for different GIS packages are readily available, users do not risk locking themselves into a particular vendor. Price points are usually at or below $1,000. If history is any guide, GIS viewers will continue to grow increasingly powerful and easy-to-use with each new release.

Who should consider GIS viewers? They will appeal to users who are looking for significant computer mapping functionality but do not need to digitize or develop their own specialized base maps. Most GIS viewer packages have a two- or three-day learning curve and require some familiarity with database or mapping concepts. GIS viewers are especially appropriate for small businesses, consultants, and local governments.

Specialized Map Analysis Tools

A fourth segment of the GIS market consists of powerful, specialized packages such as Tactician, TransCad, and the Network Analyst and Spatial Analyst analytic features available with ArcView 3.0. These programs are designed for specific types of spatial analysis and modeling. Tactician includes relatively sophisticated procedures for identifying and analyzing trade and market areas. TransCad includes a wide variety of network analysis and transportation modeling tools. ArcView 3.0's new Network Analyst and Spatial Analyst add-ons bring to the desktop environment two of ArcInfo's most sophisticated and powerful analytical components. The former can be used for projects involving market analysis, siting, and transportation planning; the latter is most suitable for land use and environmental analysis.

Most of these programs are too specialized for everyday use by real estate professionals. Their learning curves are fairly steep, and they typically require more specialized, and potentially more expensive, data.

Full-Function GIS

The top level of the GIS market consists of full-function systems such as ArcInfo, Intergraph, and ERDAS. Initially developed to run on powerful minicomputers, full-function systems migrated to UNIX-based networked workstations in the late 1980s. More recently, they have been extended to run on high-end PC servers running Windows NT.

Full-function systems typically provide hundreds of GIS functions, including sophisticated procedures for:

- Digitizing external maps
- Editing map boundaries and features
- Integrating images
- Undertaking complicated data and spatial queries
- Measuring minimum paths
- Combining and comparing features across multiple map layers
- Calculating detailed map-based relationships
- Analyzing spatial relationships in three dimensions
- Constructing three-dimensional presentation maps

Full-function GIS can be purchased as a package or in modules, with prices ranging from a low of approximately $3,000 for a basic module to more than $25,000 for a complete system. Full-function GIS packages typically have a steep learning curve, and maintaining a full-function GIS usually requires at least one full-time specialist.

The Newest Frontier: Maps and Mapping over the Internet

The newest frontier for GIS is the Internet, specifically the World Wide Web. Digital datasets (both base maps and attribute data) are increasingly available for downloading over the Internet. Common map features and associated census data for states, counties, and census tracts are available for free; data for more detailed boundaries such as zip codes or trade areas must be purchased. Most of the data published by the U.S. Geological Survey, which includes elevations, hydrography, and major roads, is available for free over the Internet.

Interestingly, a number of Web sites now allow users to construct their own thematic maps directly over the Internet and then download, or "capture," the resulting presentation map. Some university and research sites even allow users to undertake some forms of spatial analysis.

The World Wide Web is still a nonmarket, unregulated, and ever-changing environment. New Web sites and services are being added every day. Older sites can quickly disappear, or be incorporated into new sites. Web users need to know how to use Web browsers such as Netscape Navigator and Microsoft Explorer; search engines such as Yahoo, Infoseek, Lycos, and Alta Vista; and "bookmarking" features, which find and keep track of important Web sites.

FINDING THE RIGHT REAL ESTATE DATA

Real estate professionals have always recognized the importance of gathering accurate and timely information. Geographic data are more complicated than nongeographic data insofar as they reference location as well as magnitude and/or time. Thus, having the right information for use with GIS is especially important. Real estate professionals typically make use of five types of information, each of which will be discussed here.

1. Base maps, for organizing and locating all relevant information;
2. Demand data, which describes the number of space users and their willingness to pay;
3. Supply data, which describes the location, availability, and characteristics of sites and buildings;
4. Transaction data, which describes the terms of particular deals; and
5. Miscellaneous other data, including images, environmental hazards, entitlements, etc.

Base Maps

Base maps show basic geographic features such as roads, city boundaries, and parcel lines upon which real estate market data may be located or assigned. The electronic map files, often called *digital files,* consist of hundreds, thousands, or even millions of longitude-latitude coordinates defining the locations of the geographic features. In urban areas, the base maps most commonly used are built from U.S. Census maps, which are available digitally as TIGER files.[5]

TIGER files include just about all railroads, highways, roads, and streets (with address ranges); streams and other hydrologic features; administrative boundaries such as cities, counties, towns, and zip codes; and census tracts and census block groups. County boundaries, city boundaries, and census tract boundaries are all carefully cross-referenced to 1990 census data. Most of the street and boundary maps made available by private data vendors, including all the major GIS companies, are based on TIGER files. Private vendors typically enhance TIGER data in one of two ways: by correcting miscoded addresses and road types and by adding new features not in place in 1990. TIGER files are available for every metropolitan area in the United States, and most GIS software can either read or import TIGER files.

Digital maps for locations outside metropolitan areas are available from the U.S. Geological Survey (USGS). USGS data are typically provided in two forms: as a set of elevation sample points (called *digital elevation models,* or DEMs) to be used for identifying elevation and slope and as a series of lines (called *digital line graphs,* or DLGs) indicating roads, highways, and hydrologic features. Unlike TIGER files, USGS files are not linked to census data. All DEM files and many DLG files are available for downloading free of charge over the Internet. DEM and DLG files typically must be "pre-processed" before they can be read into most GIS programs.

An increasing number of cities, counties, school districts, and public and private utilities have encoded, or are in the process of encoding, assessor parcel boundaries. (Parcel boundaries are not part of TIGER files or USGS DLG maps). Parcel maps usually include parcel lines and transportation right-of-ways such as roads and railroads. Depending on the jurisdiction, they may also include the locations of utility lines and facilities, zoning districts, and actual building footprints. Local parcel maps are typically available in AutoCAD, ArcInfo, or Intergraph file formats.

Real Estate Demand Data

Demand data show the locations and numbers of real estate users. Population and/or household statistics are analyzed to indicate the demand for residential and commercial real estate, while business and/or jobs are significant to office and industrial real estate demand. Demand data are usually available in tabular rather than map form, and can be linked to maps via feature names and reference codes. Population and household data are usually organized by county, city, town or census tract;[6]

5. The acronym TIGER stands for topologically integrated geographic encoding and referencing system.
6. Individual census records for approximately 5% of the households that responded to the 1990 census are available as part of the Public Use Micro Sample, or PUMS. Because PUMS data are not spatially referenced below the metropolitan level, they are not widely used with GIS maps.

data on business, job, and retail spending, when available, are usually organized by zip code or street address. Such data are typically collected and published by government agencies such as the U.S. Bureau of the Census, state finance departments, and metropolitan or local planning departments.

Demand data are rarely collected at the site or parcel level. They are primarily used in the description of market or trade areas and are often available free or at very low cost. Most real estate professionals use demand data to prepare thematic maps as part of a market study or retail trade area delineation.

Real Estate Supply Data

Supply data indicate the amount and location of particular types of real estate: housing units, office and industrial buildings, shopping centers, hotels, and so on. With the exception of housing data, which are part of the federal census and thus available in summary form, real estate supply data are far less available (and generally less accurate) than demand data.

Like demand data, supply data are typically available in tabular form, as lists of buildings with their addresses and characteristics. Most GIS programs can make use of address matching and geocoding procedures to convert these tabular databases to point locations on a map and then summarize them by area. Supply data may be assembled with some effort from local assessor files or local parcel files, from local real estate listings (both residential and commercial), and from third-party private vendors such as Experian, DataQuick, and Transamerica Intellitech.

Supply data, unlike demand data, are rarely comprehensive. Small properties and properties that have not been involved in recent transactions may not be included. Supply data and maps are widely used in brokerage and tenant services and in the preparation of real estate market studies.

Transaction Data

Transaction data resemble supply data, but they include information on prices, rents, and sale or lease terms. Good-quality, up-to-the-minute real estate transaction data have always been fairly hard to get because success in real estate negotiations depends, in part, on who has access to what information. Sellers withhold key information to convince buyers that they are being offered the best possible deal. Similarly, buyers withhold information to let sellers know that they could easily go elsewhere.

All this is changing. New companies and strategic alliances are springing up to make up-to-date real estate information available to as broad an audience as possible. Real estate is changing and knowing how to use and analyze information is becoming more important than just having it. This dramatic change in the availability of real estate information is concurrent with, and indeed because of, technological change.

Current transaction data are most commonly available in tabular form from brokers and third-party vendors such as CoStar, COMPS Inc., and Black's Guide. Transaction data, like supply data, must typically be address matched to street maps before they can be analyzed spatially. Transaction data are especially useful in real estate brokerage, appraisals, and market studies.

Regulatory and Entitlement Data

Most regulatory data, including local plat maps, zoning districts and codes, subdivision regulations, floodplain and wetlands maps, and endangered species inventories, are still in paper form. This makes it hard for landowners, developers, and buyers and sellers of existing property to determine precisely what is and is not allowed where. As a result, nearly every major land sale or significant property improvement has required its own survey. Fortunately, more and more jurisdictions are digitizing commonly requested regulatory information, such as flood zones and zoning district boundaries, onto GIS maps and making these maps available to the public. Some cities are even making precisely digitized parcel maps available.

A limited number of vendors have begun assembling and distributing some regulatory information (e.g., data on toxic waste sites) nationwide in GIS format. These vendors include ERIIS, EDR, and On Target Mapping.

As more regulatory data become available digitally, the work of developers, land planners, survey and engineering firms, lenders, and local government plan checkers should become easier and less time-consuming.

Image Data

Image data typically consist of satellite or aerial orthophotos.[7] Image data do not include coded or interpreted features and/or attribute data. Instead, they typically consist of contrasted or colored pixels[8] or grid cells, also known as *rasters.* Image data are especially useful in trying to locate and/or code line features such as rivers, streams, railroads, highways, and power lines. Such data can also be used to locate spatially unique and differentiated polygons such as airports and parks. At higher resolutions, image data can help to locate individual building rooftops precisely.

Image and aerial photo data are widely available in hard copy form from private vendors; image scale, resolution, and age can vary widely. Many vendors also sell digitized or scanned image data in machine-readable form. Satellite imagery with one-meter resolution (i.e., the images will show anything larger than one meter square) will be available in early 1998; updated images for any site in the world will be made available every 18 days or so (when cloud cover permits).

Almost all GIS programs and many desktop mapping programs have the capability to read and register digital image data, either as background or for tagging and interpretation. Image data are likely to be of most use to developers and site planners and to real estate appraisers, site selection specialists, and market analysts.

ORGANIZING FOR A COMPETITIVE ADVANTAGE

Organizing a business to use GIS is at least as important as the GIS applications themselves. As more businesses gain experience with GIS, new management models will emerge. Currently, public agencies are the dominant users of GIS and the organization of their GIS efforts indicate three basic models: the desktop model, the management information systems (MIS) model, and the vertical integration model. The following section explores how each of these models applies to real estate applications.

The Desktop Model

The desktop model presumes that most GIS users are individuals preparing their own presentation maps or specialized projects. Desktop GIS users typically work with their own maps and databases, which may or may not be assembled on a company-wide basis. The standardization of uses and tasks across departments is less important than flexibility within departments or offices.

The typical desktop GIS user is not a GIS specialist but an individual using GIS to enhance or perform other types of analyses. Because most users do not have the time to assemble or update detailed base maps and data sets, they usually rely on secondary data or on maps and data assembled elsewhere in the company. Desktop GIS users tend to favor standardized hardware platforms running user-friendly software with common-format databases. They employ the same hardware for GIS-related tasks as for other applications and often want to transfer or link key data across applications.

Of course, desktop GIS users are not unsophisticated and may undertake very sophisticated analyses. In fact, since they are typically engaged in nonroutine, specialized projects, desktop GIS users are often at the forefront of application development.

The primary advantages of the desktop model are its comparatively low cost and its flexibility. Cost savings can be achieved by sharing hardware between applications and avoiding the overhead necessary to support a centralized system. The use of secondary source data and maps may

7. Orthophotos, or orthogonally corrected photographs.

8. The pixel or dot resolution of a conventional 35mm photograph typically exceeds 2400 dots per inch.

also keep costs down. Since each user can tailor data sets and applications to meet his or her individual needs, the desktop model promotes flexibility.

Nevertheless, the desktop model suffers from three disadvantages. Because it is dominated by individual, not group, tasks, the desktop model does little to enhance group-wide, department-wide, and company-wide capabilities. GIS capabilities fluctuate as new employees arrive and others leave. Similarly, fundamental issues of database consistency, quality, and currency tend to fall by the wayside as individual users focus on their own applications. Finally, desktop users who access GIS infrequently can find themselves "falling off the GIS learning curve."

Among real estate professionals, the desktop model is most appropriate for small firms and/or businesses engaged in market research studies, due diligence work, development-related activities, and site selection.

The Management Information System Model

The management information system (MIS) model is the opposite of the desktop model. While the desktop model assumes a distributed computing environment, the MIS model is based on a limited number of centralized data and/or applications servers. Maps and data are centrally encoded, stored, and updated, and quality assurance protocols are strictly enforced. Under the MIS model, most operations fall into two categories: those that are frequent and routine, such as the preparation of weekly sales or leasing activity reports, and those that are performed on request. The MIS model relies on a highly trained, full-time staff that provides essential computing services across several departments or locations, but is not necessarily involved in applications development.

The MIS model has been widely criticized as too limiting for today's decentralized business operations. Since the MIS model is centrally maintained, the choice of available and supported hardware and software tends to be limited. End users must fit their requests into the central framework, rather than having the central framework respond to their specific needs. In the MIS model, bureaucracy and standardization tend to inhibit adaptability and innovation.

The MIS model does have its advantages, however. It is extremely efficient, particularly in the performance of routine tasks. A high level of hardware and software expertise can be maintained, even when employee turnover is frequent. System service can be quite good, data can be shared among different users running different applications, and overall data quality can be ensured. For businesses that maintain highly accurate GIS maps and/or databases, these are significant advantages.

The MIS model is most appropriate for brokerage and leasing firms and for business in which geographic data (precise locations) is less important than processing a large volume of tabular or transaction data. Such businesses might include real estate lenders and institutional investors, REITs, retail companies, operators of multiple shopping centers, and very large development and construction firms involved in various markets.

The Vertically Integrated Model

The vertically integrated (VI) model has features of both the desktop and MIS models. Common maps and databases are stored and maintained centrally, but they are available to individual end users, usually via a network but increasingly over the Internet, and now intranets. Regular and routine tasks are performed centrally or at designated clusters, while infrequent and special projects are undertaken on the desktop by individuals and work groups. Although a single GIS package serves as the core program for most data management functions, it may be augmented by other, more specialized packages.

The VI concept extends beyond the computer system to the way that the business itself uses information. Different departments or clusters (sometimes at different locations) may be responsible for entering and updating different types of information and for undertaking different types of analyses. Some hardware and software support is centralized, but individual work groups are encouraged to develop their own areas of expertise.

At first glance, the VI model seems to have all of the advantages of the desktop and MIS models with none of their respective disadvantages. Basic services such as data and network maintenance are performed centrally, and at the same time innovation and flexibility are encouraged among end users.

The problems associated with the VI model lie in its implementation. The cost of implementing this model tends to be higher than the cost of implementing either the desktop or MIS models. There are fewer economies of scale in software licensing and hardware purchasing than under the pure MIS model. Costly hardware and software upgrades must be undertaken on a more uniform basis. Moreover, because maps tend to be far more informationally dense than numerical data, the business electronic infrastructure may require updating. (The development of business-specific intranets has the potential to alleviate some of these problems.) In addition, GIS expertise must be developed in a number of departments rather than centrally, and the sharing of data, resources, and expertise among different departments with different goals may be contrary to the company's existing corporate culture. Public sector agencies that have tried to implement the VI model have found this last barrier to be the most difficult to overcome.

The VI model is best seen as an alternative to the MIS model, particularly among larger businesses with differentiated branch or local operations. Among real estate businesses, the VI model might appeal to national or regional development companies; REITs and institutional investors; and multi-market advisory, consulting, research, and market analysis firms.

CONCLUSION

It would be a mistake to conclude that the natural fit between real estate and GIS is an essential one. Spatial data and spatial analysis have the potential to improve real estate decision making, not to transform it. Like any new technology, GIS must demonstrate that it can add real value to the bottom line. In real estate, information is valuable if it helps you recognize buy and sell opportunities before your competitors do, keeps you from paying too much for land and structures, helps you analyze and manage different sources of risk, and allows you to provide a unique service or location or the highest quality of service at the lowest cost. Although GIS may not be absolutely essential, it certainly can help in all of these areas.

chapter 2

Commercial Applications of GIS Technology

by Gilbert H. Castle III

As cited in the preface to this text, the total value of commercial real estate in the United States was estimated in 1994 to be $2.4 trillion; the value of residential real estate was estimated at $8.7 trillion. Although the residential sector is much larger, more commercial than residential real estate professionals have invested in a GIS. Commercial transactions are typically more complex and involve more monetary risk then residential transactions. Consequently, an investment in information processing and presentation technology is more easily justified by a commercial real estate firm.

Commercial applications of GIS include property valuation, brokerage, institutional real estate, and corporate real estate, among others. As each commercial real estate use is examined, the applications and benefits of GIS, case studies, implementation considerations, and the bottom line will be discussed.

PROPERTY VALUATION

Applications and Benefits

"How much is the property worth?" This is probably the most frequently asked question in the real estate industry. Regardless of whether one is buying, selling, underwriting, occupying, or managing a property, sooner or later its value becomes an important, if not critical, piece of information.

Land and improvements traditionally have been valued in three ways: using the sales comparison approach, the income approach, and the cost approach. A fourth approach is becoming increasingly prevalent: computer-assisted mass appraisal (CAMA) and related forms of automated valuation. The relevance of GIS to each approach is discussed below.

Sales Comparison Approach

Sales comparison is the most widely used approach and arguably the one with the greatest GIS potential. The value of the property being appraised (called the *subject property*) is assumed to relate closely to the selling prices of similar properties within the same market area. The appraiser first selects several similar properties (*comparables* or simply *comps*) from among all the properties that have recently been sold. Since no two properties are identical, the appraiser must adjust the selling price of each comparable to account for differences between the subject and the comparable—i.e., differences in size, age, quality of construction, selling date, surrounding neighborhood, etc. The appraiser infers the current value of the subject from the adjusted sales prices of the comparables.

The sales comparison approach is heavily dependent on the availability, accuracy, completeness, and timeliness of sale transaction data. Information sources include government records, data vendors, and the appraiser's network of local contacts (e.g., brokers participating in transactions).

The major benefits of GIS in the sales comparison approach are set forth below.

Seeing the locations of the subject and comparables on a map A comparable property does not have to be within a few hundred yards of the subject to be relevant to the appraisal, but proximity is a key factor. When many potential comparables are involved, creating a GIS display of their locations allows the appraiser to select a subset of nearby comparables efficiently for in-depth analysis. If the GIS allows the appraiser to click on a comparable on the map and display an on-screen listing of the available data on the property, so much the better. If the appraiser can click on the comparable and display photographs of its interior, exterior, and floor plan, he or she is in an excellent position to select comparables efficiently and reliably.

Obtaining supplemental information on comparable properties and the subject from other GIS data layers Traditional sources of comparable data do not typically provide information such as the socioeconomic characteristics of the surrounding neighborhood, traffic volumes on adjacent streets, and the environmental risks from a location in a flood zone or on an earthquake fault. A GIS can easily accommodate all these layers of information and quickly relate them to the locations of the subject and comparables. Using GIS, an appraiser can efficiently consider a broader spectrum of relevant variables and make appropriate adjustments to the sale prices of the comparables.

Integrating GIS with financial models, desktop publishing, and other end product tools The final appraisal report document typically contains a comparable adjustment matrix or grid that arrays and compares all the attributes of the comparables against the subject. The report also contains a map showing the locations of the subject and the comparables. With GIS products now on the market, the appraiser can click on several comparable properties on the map and the data on those comparables will automatically flow into an adjustment grid. The program then prompts the appraiser to enter adjustment factors into the grid to arrive at an estimate of the subject's value. By clicking on different comparables, the appraiser can quickly run sensitivity analyses. Once the appraiser arrives at a final value estimate, the GIS can immediately generate the requisite map of the subject and comparable properties, in color or black and white, to any scale and size. These the maps can be modified with whatever aesthetic enhancements the appraiser wants.

In short, as in any GIS real estate application, the appraiser enjoys substantial benefits by employing the GIS for analysis and decision support as well as for the preparation of presentation graphics.

Income Approach

The value of a commercial property is a function of the income it generates; the greater the income, the higher the value. An appraiser can use one or several different capitalization formulas in the income approach. A coefficient may be applied to a single year's income to calculate a value, as in the derivation of a gross rent multiplier or the capitalization of the first year's net operating income *(NOI)*. Alternatively, the appraiser can forecast income over several years and calculate the present value of the income stream, as in a five-year discounted cash flow (DCF) analysis. Innumerable variations exist depending on how "income" is defined, how many years of income are considered, whether the proceeds of an assumed sale of the property in the future are considered, and so on.

Like the sales comparison approach, the income approach can theoretically be facilitated by:

- A GIS display of the locations, physical features, and financial characteristics of properties comparable to the subject
- Direct linkages between the GIS and the appraisal formulas, ensuring rapid and thorough calculation of value indications
- Generation of a map showing the subject and other properties considered in the analysis

The benefits of GIS may be somewhat "theoretical" because data on the revenues and expenses of comparable properties are very difficult to obtain. Moreover, if the subject property is operating at a low vacancy rate, its income stream can be determined from the existing rent roll and the performance of other properties may not be relevant to the analysis. GIS will only be useful when financial data on numerous properties can be obtained (e.g., by an assessor for calculating property taxes) and if significant near-term turnovers in the property's tenant mix can be anticipated.

Cost Approach

The cost approach is predicated on the assumption that no one will pay more for a property than the estimated cost of purchasing the property as is, tearing down the existing building(s), and constructing new facilities of equal utility. Thus, the approach begins with the reproduction or replacement cost of the building improvements.

At first sight, the cost approach does not appear to benefit from GIS technology. Unlike the sale prices of properties, construction costs do not vary significantly from one part of a city to another, so using a GIS to identify spatial variations would be unproductive.

However, a GIS can help identify conditions that will significantly affect construction costs. A GIS analysis might reveal, for example, that a property is near an earthquake fault and that reconstruction will necessitate expensive seismic upgrades. Without a GIS, the appraiser might not realize that the property is subject to higher costs. (This assumes that the existing building has not been scheduled for seismic upgrades because of a grandfather clause in the local building code.)

Computer-Assisted Mass Appraisal (CAMA)

Mass appraisal is the systematic appraisal of groups of properties as of a given date using standardized procedures and statistical testing methods. Computer-assisted mass appraisal (CAMA) systems provide an opportunity to increase the efficiency and technical capabilities of an assessor's office and produce more accurate and equitable valuations. Although property tax assessors are the principal users of CAMA methods, other organizations with large portfolios of properties can also benefit from CAMA—e.g., insurance companies, REITs, and the mortgage divisions of banks.

Basically, a CAMA system builds a predictive model of property values by statistically correlating the selling prices of a large sample of properties to their physical and locational characteristics. Home values, for example, are typically correlated by the number of bedrooms, number of bathrooms, year built, architectural style, size of lot, and year sold. The correlation coefficients are then applied to the characteristics of homes that have not been sold recently to estimate their probable values.

Predictive models traditionally have been developed with standard statistical techniques, but other pattern recognition tools are being tested and implemented. In the field of artificial intelligence, for example, neural nets and case-based reasoning (CBR) automatically "learn" from the data on property sales provided. Increasing the number of sample property sales and the information available on each increases the model's reliability in predicting values for properties that have not been sold.

A GIS can facilitate computer-assisted mass appraisal in two important ways:

1. By contributing variables to the development of the predictive model and overlaying property locations on potentially important data layers, such as school district boundaries, viewsheds, distances to employment centers and retail districts, flood zones, etc.
2. By plotting the locations of properties with known sale prices along with the locations of properties with predicted values, color-coding the properties by type, and visually check-

ing for anomalies—e.g., a property with a low predicted value surrounded by several properties that have recently sold for substantially higher prices.

As CAMA and other automated appraisal methods become increasingly sophisticated and reliable, especially with assistance of GIS technology, a greater percentage of appraisals will be prepared in this way rather than via the three traditional approaches to value.

Case Studies

Given its relevance to valuation, one might expect widespread use of GIS technology in real property valuation. To date, however, case study applications of GIS in real estate appraisal have been few and limited in scope. Valuable case studies are reported in Chapters 4, 5, and 6 of this book.

One reason for the dearth of valuation case studies is the expense of investigating, purchasing, and maintaining a GIS. Most appraisers who focus on residential properties earn a few hundred dollars per appraisal and are very cost conscious. Their practices may not justify the expense of GIS technology. Commercial appraisers, tax assessors, and other valuation professionals usually have greater resources, but they too may be disinclined to try new methods without a very compelling reason.

Compelling reasons for GIS technology are emerging, however. They include the criticism the appraisal profession suffered during the savings and loan crisis of the late 1980s and the increasing popularity of CAMA and other automated valuation methods. Figure 2.1 shows an inexpensively created base map of Framingham, Massachusetts, and a thematic map of assessor's data on the annual income per square foot for each commercial property in the town. Such maps could be extremely useful to a local assessor.

Figure 2.1

Town of Framingham, Mass. **Commercial Property Annual Income/Sq. Ft.**

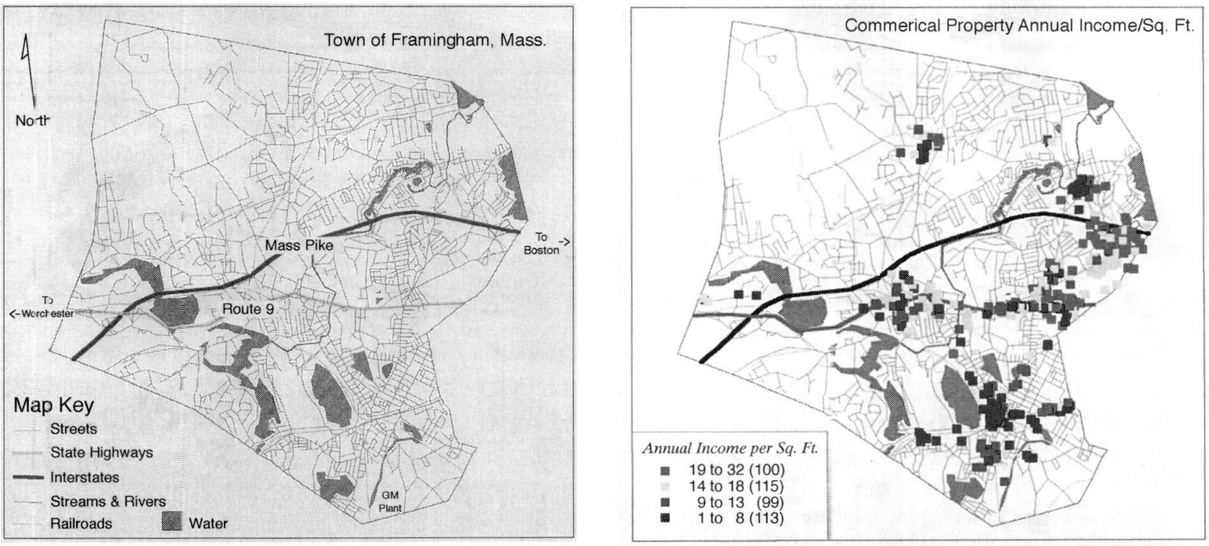

A color copy of this figure can be found on page 183.

Implementation Considerations

GIS systems can be assembled in an infinite variety of ways once a number of key decisions have been made. First, the firm has to select system components from among the dozens of software products, hundreds of hardware components, and thousands of databases available. Next, the company must decide what linkages, if any, should exist between the GIS and other computer systems (e.g., an existing local area network), databases (the organization's current management information system, or MIS), and other types of software (spreadsheets, word processing, financial models, etc.). To guide all these choices, decisions must be made concerning how the GIS will be used, by whom, when, where, and why.

Despite the diversity and complexity of GIS options, certain common concerns can be anticipated as more and more valuation professionals implement GIS. These concerns relate to job automation, data importing and address matching, and real-time linkages to non-GIS software.

Job Automation

Most appraisers, assessors, and other valuation professionals have a strict routine for doing their work. They do this to maximize the number of valuations they can perform in a given time period and to ensure that their value estimates are defensible. Accordingly, they will not flock to GIS until

- The menu options are very easy to learn and expressed in their vocabulary (e.g., sales comparison approach).
- Built-in macros allow them to compile their reports with a few keystrokes.
- The GIS-based valuation procedures they perform are virtually identical to their pre-GIS procedures.
- The GIS outputs—text, numbers, maps—plug directly into standard appraisal forms.
- Overall, they are essentially doing what they did before, but in a way that is faster, cheaper, and/or more credible.

Data Importing and Address Matching

The estimated value of a property changes the instant that a similar property nearby is sold. Similarly, values change when the rents, operating expenses, and construction costs of nearby properties increase or decrease. A valuation professional therefore must be able to obtain and analyze new data constantly. A GIS will be useful only if the new data can be entered quickly and easily. Broadly speaking, the GIS will have to accommodate the situations described below.

- Sometimes information is available only in hard copy form, e.g., in periodicals published by a local data vendor or in the notes of an appraiser's telephone calls to local brokers. Thus, the GIS will need to have preformated data entry screens for rapid input by the appraiser or clerical personnel.
- Other data are available electronically. For example, the local data vendor might publish information on recent sales not only in hard copy form but also on floppy disk in dBase format. The GIS must be able to read and process data in a wide variety of spreadsheet, database manager, word processing, and other formats. Additional complications can arise if the GIS and the data sources use different operating systems (UNIX, DOS, Windows, Macintosh System 7.x, etc.).
- Whether the data are entered manually or electronically, most data will need to be geocoded. Each property's address may be known, but not its longitude-latitude coordinate, state plane coordinate, or other earth-based coordinate. A property can be geocoded either by digitizing or address matching. Digitizing requires expert training and specialized equipment, but most real estate professionals will never need to use this procedure. Address matching is an automated process supported by most robust GIS packages; it determines the longitude-latitude coordinates of properties from their street addresses.
- A GIS will need to read as many data formats as possible quickly and easily. A considerable amount of relevant socioeconomic, environmental, regulatory, and even building-specific information is already available in geocoded format. Like textual and numerical data, however, geocoded information is published in a variety of data formats.

Real-Time Linkages to Non-GIS Software

Given that GIS-based valuation procedures should be virtually identical to the procedures appraisers have traditionally used, valuation professionals will want to link the GIS to the financial models they already use. These models may have been designed and programmed by the valuation professional or by a vendor of appraisal software. Similarly, linking the GIS to desktop publishing software will be desirable so that a GIS map of the subject and the comparables could be automatically, cosmetically enhanced and posted to a specific page in the final appraisal document. Rapid advances are being made in creating real-time links between diverse software packages, notably in a Windows environment.

The Bottom Line

The valuation of real estate is a critical activity in both the private and public sectors. GIS software and data can substantially enhance the efficiency and reliability of traditional and emerging valuation methods. The reluctance of valuation professionals to adopt new technology must inevitably give way to the obvious benefits of GIS.

COMMERCIAL BROKERAGE

Applications and Benefits

Real estate brokers facilitate transactions between buyers and sellers. The transactions can encompass all the rights to a property (fee simple ownership) or a portion of the rights (e.g., leasing one floor of an office building under carefully specified terms and conditions). A broker's compensation is usually a percentage of the amount received by the seller.

Brokers need to know who wants property of a certain type and who has property available, currently or in the near future. They should be familiar with prevailing market conditions vis-a-vis asking rents, concessions, tenant improvement allowances, and other transaction-related factors. In addition to general market conditions, brokers need to know the unique characteristics of every property they deal with.

In fact, knowing the market is not enough—brokers must demonstrate what they know. Brokerage is a highly competitive field with high potential earnings and low barriers to entry. To prosper, brokers must have both an excellent story to tell and a superb way of telling it. GIS is a wonderful tool to help appraisers "sell the sizzle as well as the steak." Using GIS technology, a broker can assemble, integrate, and analyze any available information on a property (the substance, or "steak") and displays the information with maps and other visually compelling presentation graphics (the "sizzle").

The number of brokers currently investigating and using the technology is a testament to the potential benefits of GIS. Although no empirical data are available, substantial anecdotal evidence indicates that brokers are the industry group showing the greatest interest in GIS. Their interest is most impressive insofar as brokers historically have been reluctant to spend money, due to their fluctuating incomes, and have avoided computers, relying instead on long-standing personal relationships.

It is especially noteworthy that several of the leading brokerage companies and national alliances of independent firms view GIS as a tool for significantly expanding their market share, in the near-term and long-term, both domestically and internationally. These organizations realize that economies around the globe are moving into a new age of knowledge. Brokerage firms with a reputation for having comprehensive, timely, and accurate information on real estate market conditions worldwide will enjoy a remarkable competitive advantage. What better way is there for a firm to demonstrate its knowledge than to adopt GIS technology?

As the major players jockey for competitive position, smaller brokerage companies take note. Even one- and two-person firms are investigating GIS in an effort to stay competitive. Though some are attracted to the "carrot" of cutting-edge technology, most are motivated by the "stick," realizing they need to adopt the tools their numerous and aggressive competitors are using. The growth rate of GIS in the brokerage community will likely be geometric.

How do commercial brokers benefit from GIS? Initial applications frequently relate to site selection for retail and warehouse properties. This is not surprising since GIS is well suited to the geographic and socioeconomic analyses fundamental to site selection decisions. In addition, many retailers already use GIS internally to locate and monitor their outlets, so brokers might need GIS capabilities just to keep pace with their clients.

A broad spectrum of less obvious GIS applications should prove beneficial to commercial brokers over time. These applications are described below under four general categories: marketing and sales, ongoing operations, ancillary businesses, and internal administration.

Marketing and Sales

Since one principal objective of a brokerage firm is to bring in new business, the marketing and sales staff might use GIS applications to demonstrate market knowledge and prepare exciting presentation graphics. As part of a general publicity effort, a firm might

- Employ a variety of GIS graphics in firm brochures, newsletters, annual reports, speeches, conference presentations, newspaper articles, and magazine advertisements.
- Conduct periodic, half-day promotional seminars on the firm's GIS capabilities.
- Create and publicize a "war room" for brokers, with state-of-the-art GIS and multimedia analysis and display capabilities.

To reach a targeted audience, the firm could

- Prepare prospectuses for current and potential clients in which GIS exhibits are used to demonstrate the comprehensiveness and utility of the firm's databases, analysis tools, and presentation capabilities.
- Conduct complimentary analyses of how existing clients' holdings will fare in changing marketing conditions. Such analyses might motivate these clients to buy or sell assets.
- Develop GIS-intensive promotional materials focused on creating or expanding nationwide relationships with broker networks, banks, multiple listing services, asset and property managers, investment managers, REITs, insurance companies, wealthy U.S. and foreign investors, data vendors, real estate consultants, and professional associates such as CCIMs, CREs, and MAIs.
- Publish prospectus information in electronic form for more efficient use by investment managers and other potential clients.
- Hold training sessions for the employees of favored clients, explaining the nature and benefits of GIS and related databases. By providing informational programs, a firm can cement its relationships with clients.

Ongoing Operations

Broker firms can use GIS on a daily basis to serve their clients in numerous ways.

Finding tenants for building owners

- Investigate all pending lease turnovers within a certain radius of the client's building and generate mailing labels to send promotional materials to departing tenants. GIS can be used to find out more about these and other potential tenants.
  - Identify all potential tenants within a certain travel time of the property, not just "as the crow flies."
  - Identify all potential tenants within the same submarket, as indicated by the broker or the client on the computer screen.
  - Break down the list of potential tenants to indicate the similarity between the space they currently occupy and the client's property, for more efficient marketing.
  - Break down the list of potential tenants to indicate their similarity to the client's existing tenants and/or preferred type of tenants.

After one of these screening methods has been applied, the GIS can be set up so that the broker or the client can click on the locations of the identified prospects on screen and have textual and numerical information about the firms appear nearby.

- Use the GIS to generate predefined, eye-catching maps for insertion into promotional mailings to prospective clients. The same map can be sent to all prospects or each map can be custom tailored automatically by plotting the location of the recipient's name and current location(s).
- Produce a screen display or hard copy map of the region, county, city, submarket, and/or immediate vicinity of a client's building, showing the locations and one or two key details of recently negotiated leases. The map would be prepared for the benefit of the client but could possibly be shown to prospective tenants as well. (Note that these lease comparables can be displayed as numbers, or better yet graphic symbols of different sizes, shapes, and colors can be used to indicate variations in asking rent, tenant quality, the length of the lease, or other characteristics.) As possible options, the GIS can display
  - Only lease comparables that fulfill the client's criteria in terms of tenant type, the amount of space rented, the effective rent obtained, etc.
  - Recently negotiated leases plus the current asking rents, vacancy rates, and other key data for all available properties.
  - Property-specific information stratified by building type, age, proximity to the client's building, etc., to fine tune the client's competitive position.
  - All the property information specified above presented as a time series over several months or even years. With this information the broker and client can forecast whether asking rents and other amounts will be increasing or decreasing in the near future.
- Set up an "early warning system" on the GIS that can
  - Produce zip code zone, census tract, or block group maps showing projected changes in the labor force, household buying patterns, and other socioeconomic conditions that are likely to affect leasing activity.
  - Plot the locations of pending residential and commercial development, the city's capital improvement program, vacant parcels zoned for certain types of development, and other supply-side factors likely to affect leasing activity.
  - Sort and aggregate the turnover dates of leases to predict when a relative abundance or scarcity of potential tenants will occur, for all tenants or only certain types of tenants. If, for example, the data show that no tenants of interest to a client will have their current leases expire during 1999, the client may choose to be especially aggressive in leasing building space during the remainder of 1998.
- Develop promotional literature describing the firm's activities for distribution to developers, investment managers, relocation consultants, affiliated brokers and others outside the region who have clients that may be interested in the firm's service area.
- Create an interface to a CAD package so that floor plans can be displayed and modified on screen to satisfy a potential tenant's space needs.

Buying, selling, and/or leasing of properties

- The same capabilities used to find tenants for building owners can be used to help
  - Tenants find suitable space. GIS can be applied to find all Class A buildings with space available within a specified travel time of the tenant's current leased location.
  - Building owners price and sell their properties competitively
  - Building purchasers find and bid on properties.
- Additionally, GIS can be used to
  - Find all counties or submarkets that meet specified socioeconomic, environmental, and regulatory criteria and then, within those areas, find all available properties that meet

specified real estate criteria (e.g., amount of contiguous space, asking rent, transportation access, etc.).

- Link useful GIS capabilities to standard appraisal, return on investment (ROI), asset ranking, loan underwriting, and other financial modules displayed in adjacent windows on the computer screen.
- Facilitate perusal of databases by brokers to help them identify and court new clients.
- Facilitate joint ventures with state and local economic development agencies to share in the expenses and (brokerage) benefits of a GIS for attracting new businesses.
- Again, create a "war room" with state-of-the-art GIS and multimedia analysis and display capabilities to improve both analyses and presentations.

Ancillary Businesses

Brokerage firms can achieve greater financial stability and a higher return on their assets by diversifying into full-service real estate companies. This transformation is made possible in part through a more leveraged use of their proprietary databases. As described below, GIS software and databases can figure prominently in complementary business lines such as consulting, service bureaus, and property management.

Consulting

- Offer GIS-based site selection and relocation services at a regional, citywide, and/or submarket level to clients from both outside and within the region, drawing upon demand-side databases (e.g., population and income projections at the block group level), supply-side databases (the number and types of properties available), and environmental/regulatory databases (e.g., flood zones, toxic waste sites, zoning). Such services can be provided for a consulting fee rather than as an incentive to obtain brokerage commissions.
- Conduct periodic evaluations of the real estate holdings (equity, debt, and leasehold) of corporations to alert them when changing market conditions may warrant disposing of assets or otherwise repositioning their investments.
- Perform other "outsourcing" functions for corporate real estate clients.
- Conduct periodic GIS-based "re-underwritings" of the mortgage portfolios of banks, mortgage REITs, insurance companies, and other lenders by valuing the collateral and comparing that value to the outstanding mortgage amount. This will help the lenders guard against nonperforming loans and owned real estate.
- Establish a GIS-based appraisal department.
- Use similar valuation methods and databases to conduct property tax appeals for large property owners, most likely on a contingency fee basis.
- Charge for some or most of the activities described previously under "finding tenants for building owners" rather than providing them free.
- Charge for some or most of the activities described under "buying, selling, and/or leasing properties" rather than providing them for free.

Service bureau

- Lease or sell a customized, turnkey GIS to corporate real estate executives so they can monitor the markets in which they have real estate holdings. As part of the deal, the executives might agree to purchase periodic data updates on recent transactions, asking rents, concessions, etc.
- Sell subscriptions for submarket-level and building-specific real estate data to appraisers, assessors, insurance companies, data vendors, consultants, banks, REITs, property managers, planners, and other interested individuals, locally and nationally, in both electronic and hard copy formats.

- Create and sell customized maps or complete atlases of client-selected socioeconomic, real estate, environmental, and regulatory data to appraisers, real estate consultants, investment managers, lenders, developers, etc.
- Sell or trade digital data with engineering firms, government agencies, and other GIS users. For example, an electronic map of commercial building locations could be traded for an electronic map of flood zones.
- Rent the "war room" described previously to suitable third parties when it is not needed for in-house activities.
- Establish a data reference service to help banks, insurance companies, appraisers, and others identify any environmental problems with a specific property. This can be accomplished by using the GIS to overlay the property on maps of flood zones, toxic waste sites, transmission line easements, etc.
- Establish a GIS-driven, on-line, proprietary, commercial multiple listing service (MLS) that can be easily used by specified clients, investment managers, relocation consultants, REITs, affiliated brokers, and associated firms throughout the nation.

Property development

- Select markets and submarkets to enter or exit. GIS has many well-established capabilities for market ranking.
- Analyze and aggregate major lease turnover dates to identify optimal times for new development.
- Monitor key indicators of real estate supply and demand to identify niche opportunities for new development.
- Use similar niche-identifying capabilities to attract tenants.
- Identify serious environmental problems (e.g., location in a flood zone, proximity to a toxic waste site) before making significant investments in a site.
- Create high-quality, impressive presentation graphics to be submitted to government agencies when obtaining requisite permits.
- Use GIS graphics in promotional literature designed to attract build-to-suit clients.

All of these ideas can be applied to rehab projects as well as new construction.

Internal Administration

While many brokerage firms are exploring how GIS can attract and serve clients, few (if any) seem to recognize how GIS can improve in-house efficiencies. Top management can use GIS to monitor a firm's internal operations in several ways, most notably to track office productivity, business portfolio size and composition, and changes in the business environment.

Office productivity

- Look for patterns in the geographic concentrations of the properties available versus the deals completed. The data can be stratified by property type, size and type of client, brokers involved in each deal, etc.
- Correlate the locations of deals completed in a given time period with maps showing changes in underlying socioeconomic, environmental, or regulatory conditions (e.g., ripple effects from the construction of a mall).
- Periodically compare the revenues generated by brokers who actively use the GIS to the revenues generated by brokers who do not. Compare broker performance with the above analyses of "hot" locations.

Business portfolio size and composition

- Display current and/or time series maps showing the locations of all the firm's clients, all properties being monitored, and other components of the firm's corporate database that are essential to the firm's success.

- Display similar information on the firm's competitors to assess the firm's market share now and in the past.
- Stratify the management information system, or MIS data (by property type, type and size of tenant, etc.) on the firm and on the competition to identify niches in which the firm is rapidly gaining or losing market share.

Changing business environment

- Periodically review third-party forecasts of population growth, income levels, the size and composition of the labor force, unemployment rates, housing prices, and other demand-side factors affecting the vitality of the real estate market and map them at the city, zip code zone, and census tract levels.
- Similarly, to investigate supply-side factors, periodically review maps of pending new residential and commercial development, additions to sewer and water systems, new roadways and mass transit facilities, etc.
- Monitor the number, type, magnitude, and dates of lease turnovers by submarket and individual buildings to anticipate the timing and locations of unusually high and unusually low volumes of business activity.

Most of the applications described above have not yet been implemented. Like every new technology, GIS takes time to catch on.

Case Studies

Since 1992 CB Commercial Real Estate Group in Torrance, California, has had an aggressive program to implement GIS in its offices domestically, if not internationally. To do so, CB Commercial developed and distributed MapInfo-based customized applications, geocoded all the properties in its enterprise-wide property information management system (PIMS), and offered on-site GIS training to each of its more than 80 offices in the United States. Not surprisingly, some offices have embraced the technology more enthusiastically than others. What is surprising is the efforts some brokers have made to capitalize on the perceived value of the technology, e.g., paying a GIS expert out of their own pockets to serve as an internal service bureau.

On a much smaller scale, Pizzuti Development, Inc. in Columbus, Ohio, has been effectively using GIS to promote its commercial brokerage business. Brokers and their clients can display and analyze the locations of 2,200 office and industrial properties in a two-county area; up to 42 attributes on 12,000 tenants; and related information such as photographs of the buildings. Especially attractive to property owners is the brokers' ability to generate a search radius around a client's property, find all tenants with expiring leases within that radius, and print labels for the mailing of promotional literature to these potential tenants. Pizzuti has contemplated expanding its GIS capabilities into other markets and becoming resellers of GIS data layers.

The Walker Companies of Atlanta, Georgia, is an integrated provider of real estate and consulting services to industrial corporations. The company's services include site selection for manufacturing plants and distribution centers, the disposal of surplus industrial buildings, and management of the design and construction of production facilities and distribution centers. The firm began developing the GIS in 1987, originally to support its consulting practice. (A GIS-produced map is shown in Figure 2.2.)

GIS is now used in virtually every aspect of the The Walker Companies' real estate business, especially in four areas: target marketing, matching buyers and lessors with suitable properties, competitive analysis, and market analysis.

1. *Target marketing.* Selling large manufacturing buildings and distribution centers in smaller communities and rural areas has more to do with the community and its operating cost structure than with actual bricks and mortar. GIS provides a powerful way to analyze a community's strengths and major operating advantages relative to other areas of the

Figure 2.2
GIS-Produced Map

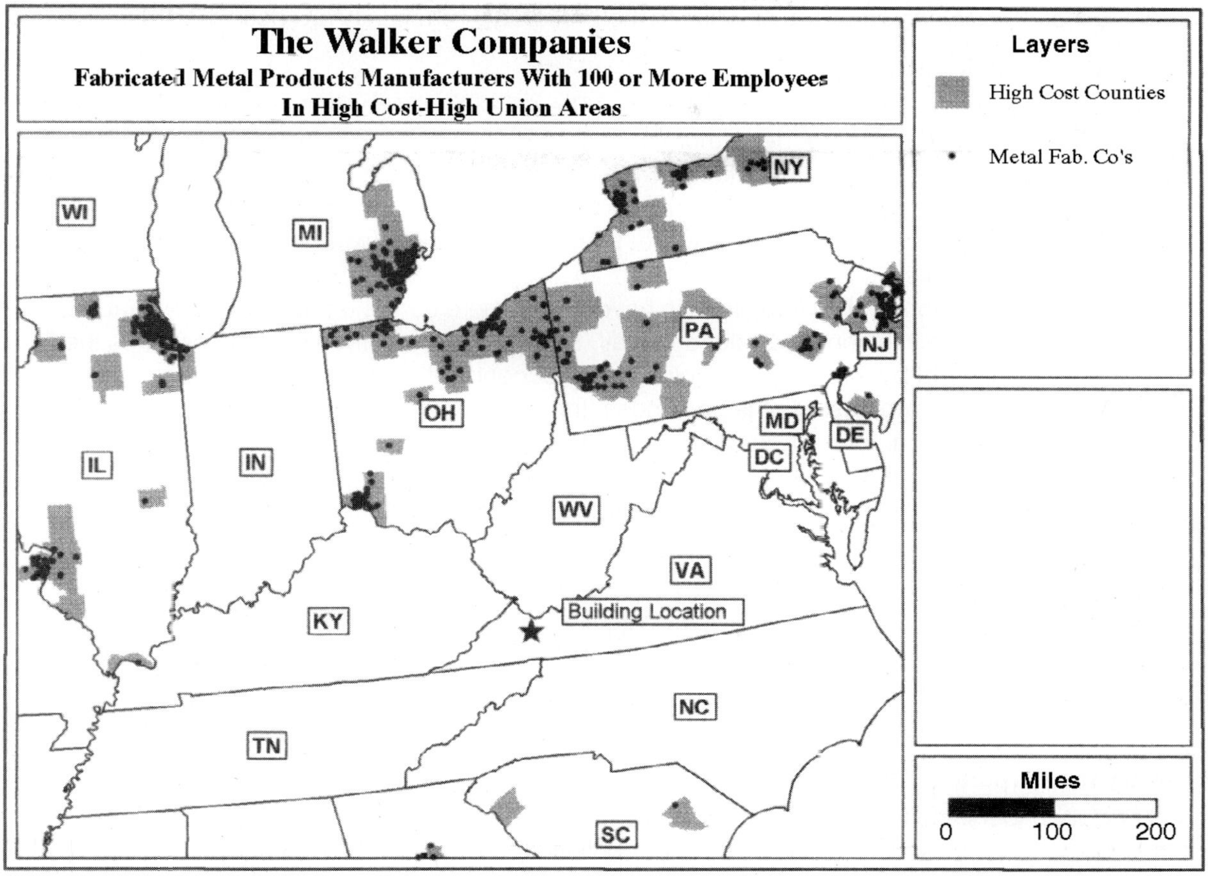

country. Then prospects that might benefit from the unique operating advantages of a particular location and building can be identified.

2. *Matching buyers/lessors with suitable properties.* The Walker Companies' databases include approximately 700 buildings that can be queried by facility size, ceiling height, the availability of utilities, rail service, distance to the closest major highway, and areas subjectively favored by the client. Outputs include not only descriptions of buildings meeting the client's criteria but also labor market profiles encompassing county-level data on manufacturing wage rates, effective property tax rates, industrial utility costs, labor union membership, union election activity, etc.
3. *Competitive analyses.* Invariably, prospects looking at a listing of The Walker Companies will also be considering properties listed by other brokers or properties for sale by their owners. The GIS allows the firm to develop a quick, yet comprehensive, analysis of all competing properties. Such an analysis also lets The Walker Companies present their listing in the most favorable light, merchandising the positive attributes of the community as well as the building itself.
4. *Market analysis.* The Walker Companies' GIS is used for a broad range of market analyses, investigating the absorption of industrial building space, changes in the demographic characteristics of a county, cost trends, operating cost comparisons, and other market factors. One of the databases contributing to these market analyses includes industrial sale and lease transactions relating to more than 1,600 properties with buildings of 50,000 square feet or more.

Implementation Considerations

The implementation considerations described earlier in this chapter in connection with real property valuation apply here as well, especially with regard to job automation and data importing and address matching. Considerations specific to the brokerage community are described below.

In-house Service Bureau

Brokers succeed by having strong "people skills." They are usually good at socializing, negotiating, and selling, and their earnings depend more on the time they spend with human beings than with computers. On the other hand, competitive brokers need the information that computers provide.

One way to solve this dilemma is to ensure that the GIS is very fast and easy to use (i.e., job automation). This solution will fall short, though, for the many brokers who are not—and do not want to be—computer literate.

Another way to solve the problem is to have one or more computer specialists available to perform computer analyses on the brokers' behalf. Such a service bureau sometimes takes the form of a junior analyst hired by three or four brokers who recognize that the analyst's salary can be paid out of their increased earnings. In other instances, the firm as a whole will set up the service bureau for the benefit of all its personnel and carry the cost as administrative overhead.

Whatever form it takes, an in-house service bureau allows the brokers to do the face-to-face work they do well and still gain access to the GIS analyses and presentation graphics they need to be competitive.

Integration with Multimedia Systems

To sell the sizzle along with the steak, a broker might use not only high-quality GIS maps but also aerial photographs, videos of the property's exterior and interior, two- and three-dimensional CAD drawings of a building's structural components, voice-overs, virtual reality walk-throughs, and any other enhancements that can help close the sale.

Mining the Corporate MIS

Brokers fully understand that real estate is a game of information arbitrage. Their success depends on assembling complete, accurate, and sometimes confidential information on as many properties, tenants, buyers, and sellers as possible. When a brokerage firm's personnel consolidate all their data into a corporate MIS that can be manipulated with powerful and flexible tools such as GIS, invaluable information can be "mined" from the depths of the database. This information can include trend data as well as current data on lease turnovers, prevailing market rents, etc. When socioeconomic, environmental, regulatory, and other GIS data layers are integrated with the real estate data, an ever-greater understanding of the real estate market can result. A brokerage firm's greatest asset might well be the corporate database that already exists in its filing cabinets and computers and within its employees' brains.

The Bottom Line

Brokers operate in an arena where the stakes are high and the competition fierce. Aggressive pragmatism (find, kill, eat) necessarily guides their acceptance of new tools and techniques. The fact that many brokers are evaluating and embracing GIS suggests that the technology is likely to be adopted by the rest of the real estate industry.

INSTITUTIONAL REAL ESTATE

Applications and Benefits

Institutional real estate encompasses pension funds, Wall Street security entities (e.g., REITs), insurance companies, wealthy U.S. and foreign investors, and others investing purely for profit—not to occupy space. Although institutional investors are a much smaller source of capital than corporations satisfying their space needs, they are considerably more visible for a number of reasons:

- Institutional investors may be the only active players at times, e.g., during a recession when corporations are postponing or scaling back their investments in space.
- Regardless of economic conditions, corporations acquire or dispose of real estate relatively infrequently; institutional investors are in and out of the marketplace all the time.
- Corporations have many criteria for space that meets their needs, so they can narrow the list of candidate properties relatively quickly and quietly. Institutional investors, on the other hand, essentially have only one criterion—a specified, risk-adjusted rate of return on capital—so there are usually many more candidate properties to identify and evaluate.
- Nonfinancial factors can generate publicity for institutional investors. For example, escalating U.S. real estate investments by the Japanese were given significant media coverage in the late 1980s.
- A major portion of the actual purchase and management of real estate assets is handled by investment management firms, Wall Street companies, and other financial conduits that must first attract investment capital from pension funds, wealthy individuals, and other investors. This is a highly competitive business and the players have to be active self-promoters.

Due to its competitive nature, the institutional real estate market is characterized by the use of highly sophisticated financial models, databases, and presentation techniques. The more sophisticated a firm is, or can represent itself to be, the greater that firm's competitive edge in attracting and placing investment capital.

This is where GIS comes in. GIS is unquestionably a sophisticated technology for assembling, analyzing, and presenting real estate data. Again, the sizzle and the steak are both essential, and GIS provides the desired benefits.

GIS has been used by institutional investors in three principal ways: market ranking, which has received the greatest attention to date, submarket analysis, and due diligence.

Market Ranking

The 50 to 100 largest metropolitan statistical areas (MSAs) in the United States have traditionally been the markets of greatest interest to institutional investors. The preferred, "core" property types have been office, retail, warehouse, and apartments. These four property types multiplied by 100 MSAs give the analyst 400 combinations to investigate. Additional combinations could also be considered given the number of submarkets within each MSA (frequently five or more per MSA) and subcategories of property types (e.g., high-rise, low-rise, and garden apartments). In fact, the investment universe can quickly grow to include 4,000 combinations of property types and geographic markets.

To avoid being overwhelmed by the data collection and analysis requirements of 400 to 4,000 combinations, many institutional investors employ some type of market ranking methodology in which each property type-market combination is assigned a score based on various real estate demand and supply factors. More detailed analyses can then be performed on the most promising combinations. The analyst using a macro filter and a micro follow-up might occasionally overlook a promising investment opportunity, but this is better than having to build and maintain huge databases.

Each institutional investor has a unique scoring model, which is believed to be better than anyone else's and is promoted as a competitive advantage. Broadly speaking, these models differ in methodology, the variables they use, and the role of GIS.

Methodology Many options are possible, including

- Simple screening, such as investigating only those markets that have experienced population growth above a certain rate and are projected to continue to do so.
- A scoring model in which each market receives a score calculated from the values for several variables, such as

$$S = (50\% \times P) + (30\% \times I) + (20\% \times E)$$

Where *S* is the overall score for a given market for a given property type
P is the average percentage change in population over five years
I is median household income normalized over all the markets
E is the current employment rate

The values are usually normalized and the variables are typically weighted.

- A pattern recognition program for correlating investment returns to market characteristics, using either classic statistical methods or new artificial intelligence techniques such as case-based reasoning or neural nets.
- A financial model in which a 10-year pro forma statement is developed for an average property in each market based on local asking rents, vacancy rates, etc. Then the return on investment (ROI) for each average property is calculated from the pro forma using standard formulas (e.g., internal rate of return, cash on cash, pay-back period), and the markets are ranked from best to worst based on one or more of the ROI figures.

Variables Some investors are "demand-siders," who believe that no viable financial opportunity can exist without significant growth in population, employment, income, and similar socioeconomic factors that correlate to demand for new stores, offices, apartments, etc. Other investors argue that demand is important—but so is the supply of real estate. Too often, cities with strong demand for new offices, stores, and other enterprises experience excessive construction, and property values fall. Moving past the demand versus supply debate, reasonable people can differ on how much reliance should be placed on:

- Historic versus current versus forecast figures.
- Changes measured in absolute numbers as opposed to percentages.
- Whether to weight different variables and by how much.
- Extrapolating a single, most likely scenario or several scenarios, e.g., optimistic, realistic, pessimistic, etc.

Role of GIS Since the geographic units of analysis are 50 to 100 MSAs, institutional investors only need rudimentary GIS presentation capabilities to do market ranking. The typical GIS display will be a map of the United States showing state boundaries, possibly interstate highways or other reference features, and the boundaries of the 50 to 100 MSAs analyzed. The display will show the values of one of the ranking model's inputs or outputs by color-coding the MSAs, by using proportional symbols or three-dimensional shapes, or by employing some other presentation technique.

Investors may use a variation of market ranking called *portfolio management.* Some institutional investors believe that the highest risk-adjusted returns are obtained by diversifying real estate investments on the basis of property type, the geographic dispersion of the MSAs, the economic base of the MSAs, and other factors. Highly sophisticated models can be used to identify pairs of MSAs with negatively correlated real estate cycles. GIS displays of the United States can depict these diversification factors.

Submarket Analysis

Submarket analysis, also called *market research* or *market monitoring,* focuses on one or more submarkets within an MSA. It can be applied as a more detailed, follow-up step to a market ranking project, i.e., conducting in-depth analyses of the five to 10 MSAs that were identified as the most promising. Alternatively, such an analysis can be undertaken to identify significant changes in submarkets where previous investments have been made and serve as an early warning system for disposing of potentially problematic assets. Thus, submarket analysis has applications in acquisition, asset management, and disposition.

The power of a GIS can be exercised far more completely in submarket analysis than in market ranking because a greater number of maps are used to indicate submarket boundaries, zip code zones, census tracts, block groups, parcel boundaries, streets, environmental hazards, and other features. Far more interesting and numerous presentation graphics can be prepared, and the power of GIS can be used to integrate different data layers. For example, the analyst's request might state:

> Display all sites (parcel layer) in a rapidly growing area (census tract layer) where unemployment is low (typically the zip code zone layer) that are within one-fourth mile of a freeway interchange (streets layer) and not in a flood zone (environmental hazards layer).

Analysis models and input variables differ from one investor to another. Regardless of the methodology of choice, however, GIS technology can be a valuable tool both in supporting business decisions and in the presentation of findings.

Due Diligence

Due diligence requires an objective, thorough, and sometimes fiduciary review of the risks and rewards of a proposed property purchase. It may include related endeavors such as project feasibility studies and highest and best use analyses. The geographic focus of analysis is usually a single submarket or the area within a certain distance (say, five miles) of a specific site. Due diligence is required for property acquisition only, not for ongoing asset management.

The use of GIS in the decision making and presentation of a due diligence study is similar to its use in submarket analysis but somewhat more restricted. The restrictions are partially caused by a smaller geographic focus—i.e., a smaller study area results in the need for fewer zip code zones, census tracts, environmental hazard maps, etc. Moreover, a due diligence study usually places greater emphasis on the financial aspects of the transaction (e.g., the creditworthiness of the buyer), which may or may not have a geographic component.

In summary, the three broad ways in which institutional investors use GIS—market ranking, submarket analysis, and due diligence—form a geographic hierarchy, descending from a national to a local to a property-specific level of analysis.

Case Studies

Most, if not all, of the 35 or so leading institutional investment management firms have a GIS. Early adopters include Equitable Real Estate Investment Management in Atlanta, RREEF Funds in San Francisco, and Aldrich, Eastman & Waltch in Boston.

Perhaps the most comprehensive and sophisticated applications of GIS to date, however, have been performed by State Street Research Realty Advisors (SSRRA) in White Plains, New York, formerly the MetLife Realty Group (MRG). As early as 1995, geographic technology was identified as the best tool to help SSRRA's analysts gather, sort, analyze, and display potential risk-return data and show spatial relationships among real estate supply, demand, and regulatory factors. The ArcView-based prototype system developed was scripted for each department within SSRRA—i.e., the log-on screen displayed buttons labeled "Acquisitions," "Asset Management," "Appraisal," "Library," and so on to call up GIS-based maps and reports tailored to the individuals using them. (See Figure 2.3.) The prototype confirmed that GIS technology can

- Be easily used by everyone in the organization.
- Efficiently integrate all the data relevant to real estate decisions into one system.
- Generate compelling presentations for telling the SSRRA story.

Top management subsequently authorized the development of a full-blown system. The evolving system is migrating from ArcView to MapObjects; both are software products of ESRI in Redlands, California.

As Figure 2.4 shows, SSRRA brings together extensive data sets from third-party vendors and proprietary, in-house databases. Oracle is the principal data "melting pot," with GIS sitting on top for accessing, analyzing, and viewing the data. In addition to GIS analyses, the Oracle database fuels a broad spectrum of special-purpose software programs, such as linear programming modules, simulation tools, and spreadsheets.

Figure 2.3
GIS Maps Tailored to Users

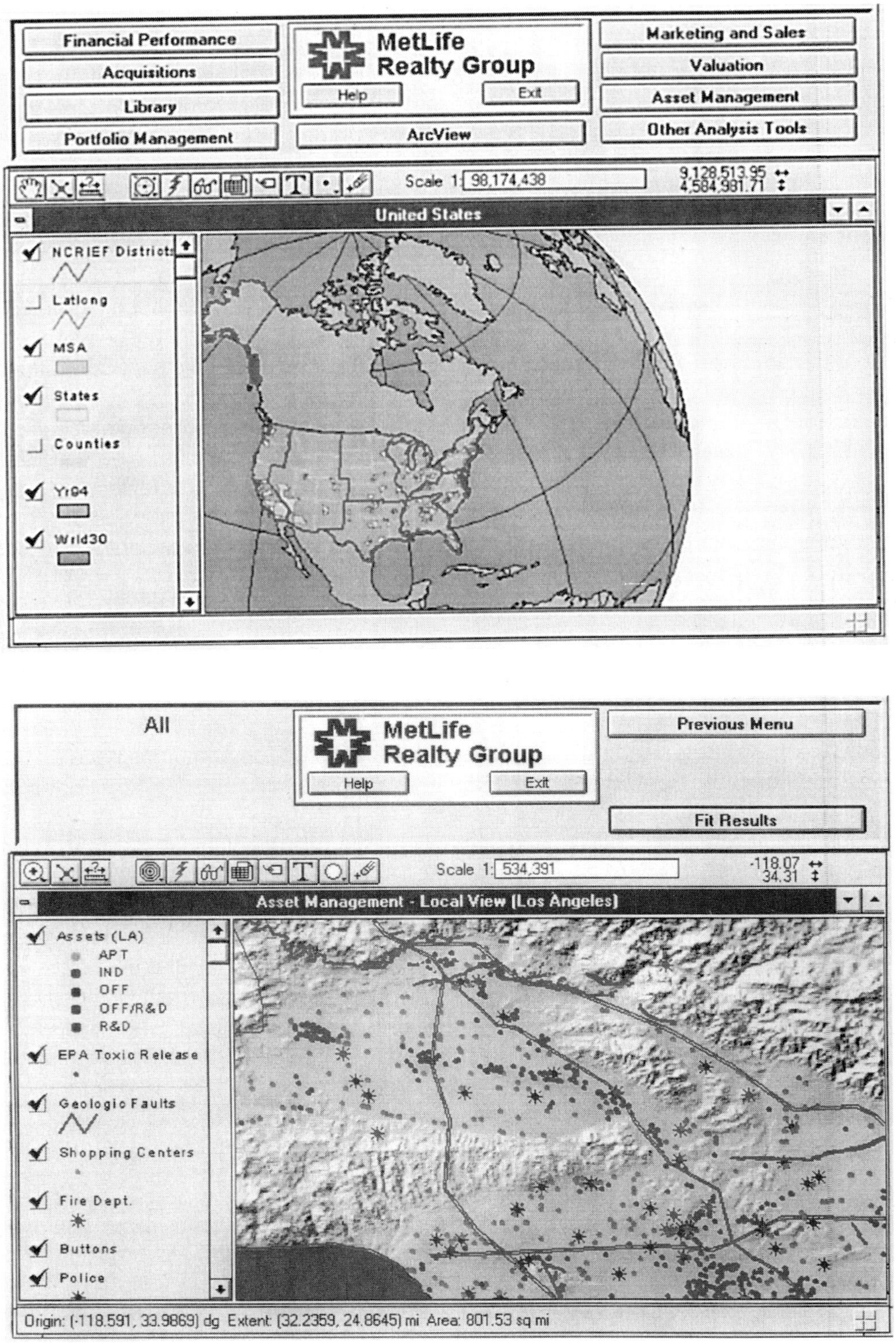

A color copy of this figure can be found on page 184.

Figure 2.4
GIS-Based System Diagram

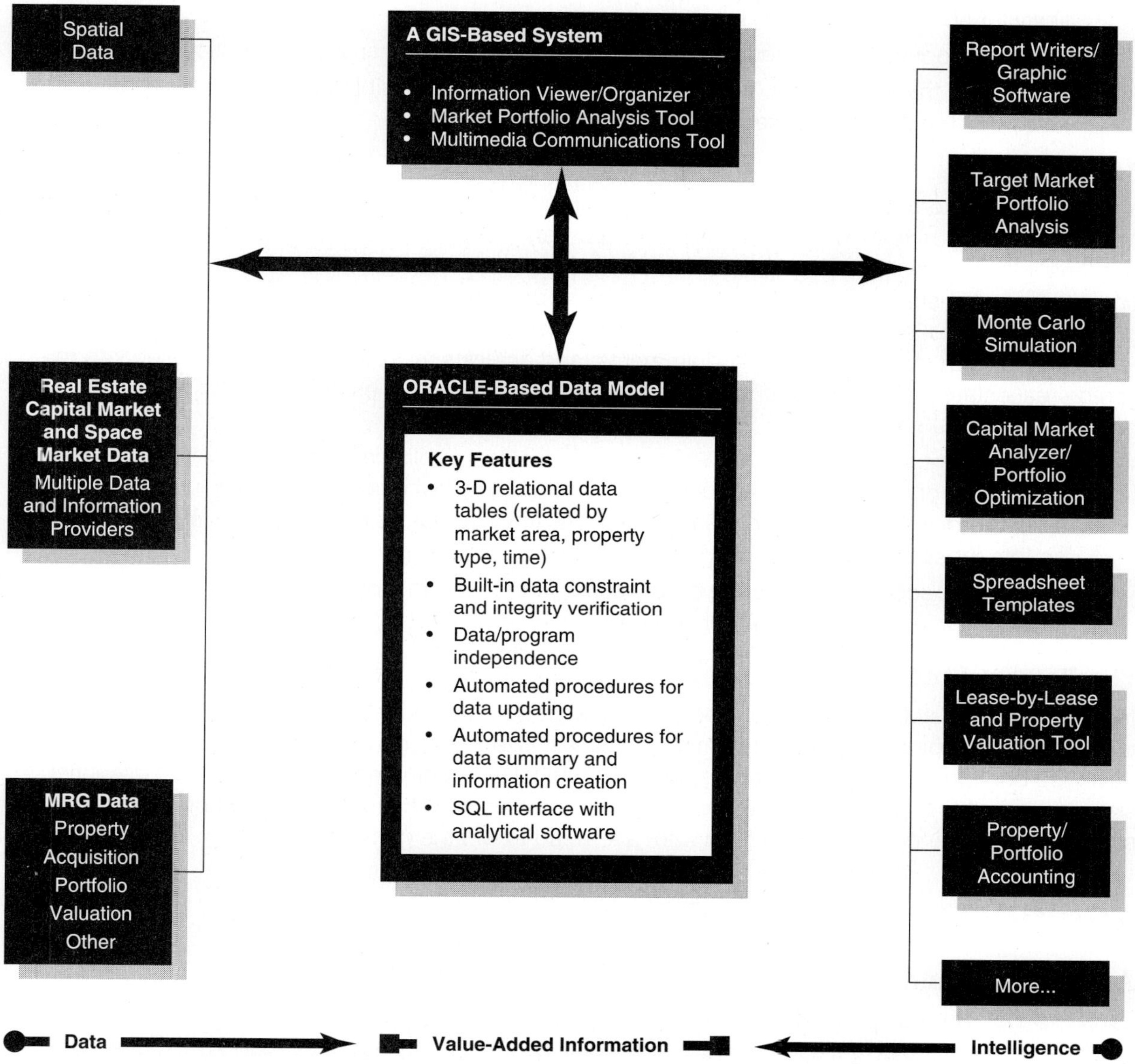

In the highly competitive and cerebral world of institutional real estate investment, SSRRA obviously believes that GIS provides a significant business advantage.

Implementation Considerations

Institutional investors are concerned about many of the implementation considerations cited previously, but they place the greatest emphasis on reliability and accountability.

Reliability

Institutional investors want to know that their real estate investments will achieve a target rate of return, the "hurdle rate," and that the value of the asset will not diminish significantly. This is especially true when fiduciary responsibilities are involved, as with pension funds. Accordingly, institutional-grade properties are high-quality, conservative, well-researched investments. A GIS implemented to find, acquire, and manage institutional investments must therefore

- Demonstrate sophisticated capabilities for assembling, analyzing, and presenting information crucial to an investment decision.
- Encompass extensive databases at the national, local, and property-specific levels, most likely for 50 to 100 MSAs.
- Support proprietary methodologies for market ranking and other crucial functions.
- Establish a track record for "picking winners" as soon as possible to assure institutional investors that their funds are being invested wisely.

Accountability

Institutional investors do not usually hand over their funds and walk away. They want to be updated, usually quarterly, on how well their investments are performing. The reporting requirements can be burdensome, especially for organizations with fiduciary responsibilities. Quarterly reports typically contain current, historic, and projected returns by property type, location, investment manager, etc., presented in rigorously defined tables, charts, and maps. Assets that are not achieving expected returns have even stricter reporting requirements, and analysts must often provide detailed information on real estate market conditions in the vicinities of those properties. In support of these audit requirements, a GIS must

- Complement the database management and desktop publishing process in which the most recent rate of return figures are combined and analyzed with figures for previous quarters and a standardized document is generated as efficiently as possible, probably using macros or scripts (possibly integrating the GIS with desktop publishing).
- In addition, accommodate ad hoc queries for further information on problem properties and other nonstandard reporting requirements.

The Bottom Line

Institutional real estate is a game played for very large sums of money, under bright spotlights, with innumerable competitors watching from the sidelines. Clients are demanding, always asking, "What have you done for me lately?" Having a good story to tell and being able to tell it well are essential skills. An increasing number of institutional investors are demonstrating that GIS embodies the sophisticated science and artistry they want and need to win the game.

CORPORATE REAL ESTATE

Applications and Benefits

Corporations needing space for their own use are the source of approximately 75% of the capital flowing into commercial real estate. Real estate holdings account for approximately 25% of the average corporation's assets. A company's annual expenditure on real estate is typically two times its net earnings. Corporate real estate is by far the largest component of the commercial real estate industry.

GIS relates to corporate real estate both directly and indirectly. The indirect relevance takes the form of GIS use by organizations serving the corporate real estate market, including brokers, appraisers, developers, underwriters, and others, as discussed in other sections of this chapter. The direct relevance is evidenced by corporations using GIS in-house for acquiring, managing, and disposing of real estate.

Traditionally, a corporation's real estate department has been viewed as a minor part of the business, a cost center that does not even need the supervision of a full-time executive. The only significant decisions to be made—selecting sites for new facilities or relocating existing facilities—were infrequent and typically handled as special cases.

In recent years, however, corporations have begun to realize that site selection and relocation are not the only real estate decisions that affect the bottom line. During the heyday of mergers and acquisitions, takeover specialists frequently would find hidden value in a targeted company's real estate holdings. These holdings might be carried on the company books at the original purchase price or at an outdated market value, rather than at their current market value, which might be substantially higher. Then and now, viable businesses are liquidated because the value of the real estate exceeds the going concern value of the operations. In the current climate of stiff competition, layoffs at all levels, and a return to business basics, CEOs and CFOs are searching diligently for opportunities to increase revenues and decrease costs from all corporate assets—including real estate that is owned, leased, or otherwise controlled. Ongoing asset management is the key.

Broadly speaking, GIS complements ongoing asset management in five ways: property management, rightsizing, repositioning, outsourcing, and property tax abatement.

Property Management

Asset management usually refers to a portfolio of properties, while property management refers to a specific facility. The most obvious responsibility of a corporate real estate manager is to ensure that appropriate space is being provided for company operations at a reasonable, if not bargain, price. If the space is being leased, the manager has to know about prevailing asking rents, concessions such as a month or more of free rent, dollar allowances for interior improvements, options for lease renewal, and so on. If the corporation owns the facility, the manager has an even more comprehensive and complicated job: he or she must not only maintain smooth and efficient functioning of all services (e.g., utilities, janitorial, security) but also strive to acquire those services at or below prevailing market rates.

The key to success is having timely and accurate data, which the manager can then use to negotiate the most favorable terms for leasing space, hiring a janitorial service, etc. A GIS is an excellent vehicle for organizing, retrieving, and displaying information. A corporate real estate manager can implement a GIS showing all the properties within the city that are similar to the one that he or she is administering. For each of these properties, the manager can acquire and input the latest available data on various categories of revenues and expenses. The GIS might also include photographs of each property's exterior, interior, and floor plan; regulatory information, such as the creation of a tax increment financing zone that will affect some properties but not others; environmental maps that may reveal newly discovered earthquake faults, necessitating seismic upgrades in nearby buildings; and other data.

With a GIS the manager can

- Create a thematic map in which every property is coded with a color and/or symbol to indicate physical features (e.g., total square footage, year built), financial characteristics (e.g., vacancy rate), tenant mix (e.g., one, few, or many tenants), and other characteristics.
- Display all properties fulfilling certain criteria of interest, e.g., one or more major lease turnovers within the last six months.
- Click on selected properties to identify differences in the rental rates of the old and new leases and thus determine upward and downward trends in the market.
- Download or directly link data from the other properties into non-GIS financial models to calculate potential changes in the manager's monthly pro forma.

Obtaining the data to support a property management GIS is not a trivial undertaking. Information such as lease turnover dates are jealously guarded by brokers and property managers. Still, more and more data sources are emerging, including new types of real estate data vendors and data exchange cooperatives. Given the dollars at stake in even a single lease, the resources and managerial effort needed to build and maintain a GIS database can most likely be amortized quickly.

Rightsizing

Most companies need to ask

- How much much space does the corporation need to support its operations efficiently, now and in the foreseeable future?
- Where should that space be located?

- How do current and future space needs compare to the amount and spatial distribution of the real estate the corporation owns, leases, or otherwise controls?

Ensuring an efficient match between the demand for and supply of space for business activities is called *rightsizing.*

Rightsizing is sometimes considered a euphemism for downsizing. Recession, global competition, corporate re-engineering, reinventing government, and other economic pressures continue to result in staff reductions. Fewer workers mean less space is needed.

In a parallel development, corporations are exploring ways of reducing the amount of space needed per worker. Business publications report that companies are experimenting with "virtual offices" and "hoteling" to eliminate the need for employee offices. The virtual office concept capitalizes on technological advances in communications equipment and computing, which allow employees to work out of their homes or even their briefcases. In hoteling, employees call ahead to reserve a temporary office or cubicle at corporate headquarters when they are not working at a client's site. Under either plan, less office space is needed.

As the demand for space diminishes, the corporate real estate manager can consolidate real estate holdings into more efficient layouts and dispose of surplus space. Disposition might take the form of an outright sale of a building or leasehold interest or the releasing of space.

The financial results of rightsizing can quickly justify the effort. Since Class A office space in most cities costs $25 to $45 per square foot per year, and the average office worker uses 150 to 300 square feet (including common areas), the savings for each employee can range from $3,750 to $13,500.

A GIS can help the corporate real estate manager capture those dollars in numerous ways, by improving asset inventory, market monitoring, and employee efficiency and morale.

Asset inventory An amazingly large number of both private and public sector organizations do not have a central, computerized database on all their real estate holdings. Very few currently have such a database linked to a GIS. A simple but incredibly powerful use of a GIS is to input the locations, physical features, and financial characteristics of all real estate holdings and then display that information on one or more maps to help decision makers visually identify opportunities for space consolidation and disposition.

Market monitoring Some types of surplus space are easier to sell or sublease than others. To optimize disposition revenues, the corporate real estate manager will need to keep track of the amount, distribution, and asking price of competing space already on the market. The manager can accomplish this task using the database and procedures described in the previous section on property management.

Employee efficiency and morale By entering into the GIS the locations of the company's major clients and residential addresses of employees, the corporate real estate manager can make well-informed space consolidation and disposition decisions. All else being equal, the manager should retain corporate space that minimizes employees' travel time between their homes, the office, and clients. Some GIS products will automatically map these locations for visual analysis and also calculate straight-line distances and actual travel times among numerous sites.

Though the focus of this discussion is office space, the same siting procedures can be applied to factories, warehouses, retail outlets, and other property types.

Repositioning

The location and amount of space in a given real estate asset may be right, but its financial characteristics may not be. In such a case, the corporate real estate manager might be able to transform his or her department from a cost center into a profit center by recapitalizing one or more problematic assets. By way of example, consider the following scenarios:

- A company currently owns the building it occupies. Reliable real estate market forecasts show that commercial values will soon decline significantly due to an excess of office space under construction. The corporate real estate manager sells the building while the value is still high and structures the deal so that the company can remain in the building at favorable terms for a year. The following year the manager either signs a new lease for the same space or moves to a new building, taking advantage of the lower market rents.

- Another company is leasing space in a market that is projected to enjoy increasing commercial values. The company's core business is expected to experience a slump for one or two years, due to an anticipated temporary increase in manufacturing costs (say, a short-term increase in the cost of energy to power the factories). Working closely with the company's CFO to find and reinvest surplus capital, the corporate real estate manager identified opportunities to convert leases into equity positions. As the market value of these newly purchased properties increases, the company enjoys a greater overall return on assets (ROA) than would have been possible if the company had continued to lease, rather than invest in the space it occupies.
- A third company has factories that are well located, but the manufacturing equipment is outdated and must be replaced. The company is reluctant to borrow the funds needed for the upgrades because current interest rates are high. On advice from the corporate real estate manager, the company discovers that it can raise the needed capital through a sale-leaseback. The company sells its factories to one or more third parties, retaining the right to remain in the space. The cash generated by the sale is used to pay for the new machinery needed and funds are left over for other corporate uses. The efficiencies of the new machinery reduce production costs more than enough to cover the lease payments.

The essence of repositioning is continually looking for ways to increase the ROA from the company's real estate holdings just as the company looks for ways to increase ROA by investing in new or different factories, personnel, subsidiaries, or other assets.

GIS supports repositioning by providing current and forecast information on the real estate market. The databases involved would be similar or even identical to those used for property management and rightsizing. Supplementing these databases would be periodic (quarterly or semiannual) appraisals of all owned properties and leasehold interests. By monitoring changes in the properties' appraised values, the corporate real estate manager has an early warning system for detecting upward or downward market trends. Moreover, having current information on the values of all the company's real estate holdings is essential if the corporate real estate manager and the CFO are to be successful in their search for repositioning opportunities.

Incidentally, GIS also supports the repositioning process by facilitating cost-effective appraisals. Although a corporate real estate manager might not be able to justify the significant expense of retaining a certified appraiser to value each property every quarter, it might be cost-effective to implement an in-house, GIS-based CAMA valuation system or to value properties using a GIS with the sales comparison approach.

Property Tax Abatement

In the past decade, property values have declined in many parts of the nation. Property tax assessors are not always as prompt or diligent as corporations would like in lowering property taxes in step with diminishing market values. Corporations can sometimes save hundreds of thousands of dollars by appealing their property tax assessments.

As described in the real property valuation section at the start of this chapter, GIS can be used effectively in all three traditional valuation approaches. The corporate real estate manager can quickly and efficiently explore various valuation scenarios to find comparables and other relevant data in support of the tax appeal. As noted earlier, a GIS is very compatible with computer-assisted mass appraisal. In jurisdictions where the assessor is using a GIS in conjunction with a CAMA system, the corporate real estate manager should be able to acquire the assessor's valuation date—and then use the same software and data to substantiate the property tax appeal.

Many corporations retain an independent property tax consultant to conduct appeals on their behalf. These consultants sometimes work on a fixed-fee basis; more frequently, however, they work on a contingency basis, keeping a large percentage of the savings they achieve for the corporation. A number of these consultants are now investigating the use of GIS in their businesses. With GIS they can test more valuation scenarios in less time, replicate the valuation procedures of assessors who are using GIS, and/or use powerful GIS data display tools to create highly professional, persuasive presentation graphics.

Thus, directly or indirectly through a consultant, a corporate real estate manager can use GIS to reduce property taxes, with the savings flowing straight to the bottom line.

Outsourcing

To lower costs, many corporations have reduced the size of their corporate real estate departments through outsourcing. Rather than keep appraisers, brokers, and property managers on staff, a corporation might subcontract for these services with third parties. Broadly speaking, a GIS can help the corporate real estate manager ensure the cost-effectiveness of outsourcing in two ways:

1. As a reality check. Since GIS can be used in appraisal, brokerage, property management, and other real estate disciplines, a corporate real estate manager can employ GIS to replicate a few or many subcontracted analyses and verify the quality of the work. Using the valuation capabilities of a GIS, for example, the manager can look for flaws in a submitted appraisal or challenge the price that a broker wants to ask for a surplus property. In this way, the manager fulfills his or her responsibility to maintain quality control and extracts the best work possible from the subcontractors by letting them know that their work is being rigorously monitored.
2. In outsource administration. If a corporation has a large portfolio of properties distributed over several cities, states, or even nations, a corporate real estate manager can easily lose track of who is supposed to do what, by when, and where. Just as the locations and characteristics of the company's properties can be stored, retrieved, analyzed, and displayed in a GIS, so too can information on subcontractors. With a GIS database, a manager can
   - Display a map of all company properties in a given city, click on one property of interest, and have the GIS list all the outsource contracts pertaining to that property.
   - Show the locations of all properties that have been valued by a certain appraiser or all those in the process of being sold by a certain brokerage firm.
   - Display the operating expenses of all properties in an area and then identify the property management subcontractor for each to determine whether one property management firm is substantially more competent than the others.
   - Draw a circle around a property on a map and have the GIS list the names and telephone numbers of all appraisers working within that circle, as an aid in finding new outsource candidates.

The above discussion has focused on private sector organizations, but public sector agencies also control substantial amounts of real estate. Several states have investigated and implemented asset management programs (e.g., Texas, California, Hawaii). Except for the discussion of property tax abatement, everything in this section on corporate real estate applies also to the activities of government real estate managers.

Case Studies

Chevron Corporation in San Francisco, which is one of the largest companies in the world, has extensive real estate holdings. In 1988 Chevron began assembling a GIS database on all its real estate holdings as well as those of its operating subsidiaries. Today information on all of Chevron's properties in approximately 800 U.S. counties is stored in an online GIS database referenced to 1,500 United States Geological Survey (USGS) quadrangles. The system employs software from Intergraph Corporation of Huntsville, Alabama, and the spatial data are linked to Chevron's worldwide, corporate MIS, which is based on NOMAD database management system software. The real estate information is updated weekly. Queries of the data can originate in NOMAD or in the GIS. Among other uses, the system

- Generates hard copy maps of property boundaries for Chevron's in-house real estate professionals in both the parent company and the subsidiaries.
- Responds to questions from brokers and other external parties who want to know if certain parcels belong to Chevron and what the status of those parcels is.
- Provides a spatial reference (e.g., a building footprint on a local area street map) for facility managers and office space designers who are charged with refitting the interior of Chevron buildings.

The real estate GIS is used every day. Chevron executives found the system to be especially useful when disposing of surplus properties became a high priority.

Moving from traditional industries to the world of high tech, Sun Microsystems Computer Corporation of Mountain View, California, has an evolving system that combines real estate asset management and facilities management. With leases worldwide, Sun managers need simple and convenient access to real estate information at the facilities level and at the Executive Information System (EIS) level. Sun had been using CAD technology but recognized the need to move to GIS for two reasons: Sun needs to associate data with facilities and do spatial processing, e.g., find all assets "near to" something else. As a first step, Sun has geocoded 170 properties nationwide, using GIS software from ESRI in Redlands, California, and digital maps from GDT of Lebanon, New Hampshire. Data such as the lease dates and square footage of each property are stored in a SYBASE database. By clicking on a property (shown as a symbol) on the computer screen, the attribute data can be retrieved and displayed. Interior and exterior photographs can be retrieved (via Kodak Photo CD) as well as CAD drawings (via Postscript or TIFF).

Implementation Considerations

The implementation considerations discussed earlier under real estate valuation apply here as well, with a few variations relating to standardization, linkages to non-GIS software, larger databases, and team effort.

Standardization

Corporate real estate managers will need their GIS to support established policies and procedures for acquiring, administering, and disposing of real estate assets. However, those policies and procedures are likely to be more far-ranging and more frequently modified than the rigorous financial models employed by appraisers, assessors, and other valuation professionals. The corporate real estate manager's standards are also likely to be more qualitative and less quantitative, e.g., taking into account impacts on employee efficiency and morale.

Linkages to Non-GIS Software

Again, the corporate real estate manager will need greater flexibility than the valuation professional. The manager will not only want a link to standard appraisal software but to a variety of other software as well. The GIS might be linked to

- Sophisticated database management systems specifically designed to inventory and monitor all the real estate holdings of a corporation.
- Special-purpose software such as site selection programs.
- Project management software, electronic rolodexes, and even automated tickler files for administering outsource contracts.
- The company's main MIS, from which a wealth of information can be gleaned on the locations of clients, employees, etc.

Larger Databases

A valuation professional typically works in one city and collects data for just that city. In contrast, a corporate real estate manager usually oversees assets in many cities. When a company is considering expanding or relocating, the manager might need some level of information on virtually every city or county in the nation. In fact, databases might be needed for several countries, not just the United States. The manager must make challenging decisions concerning the amount and type of GIS information to collect and maintain at the national, state, local, and property-specific levels. It can be a complicated trade-off between keeping costs as low as possible and acquiring sufficient data to make sound decisions.

Team Effort

Many appraisals are prepared by one person, frequently a sole practitioner. These individuals have the luxury of setting up their GIS software and databases to suit their work style. Sometimes a corporate real estate manager can do this too, but more often the manager must instead create a flexible system that accommodates

- The numerous and diverse roles played by the manager, extending well beyond property appraisal.
- Cooperation with other departments within the company, especially when these departments retain line authority for the real estate they occupy and view the corporate real estate manager only as a staff resource.
- Administration of outsource contracts, which may require the corporate real estate manager to replicate some of the data processing capabilities of subcontractors.

In short, the corporate real estate manager's GIS software and databases usually must be more flexible and comprehensive than the systems of appraisers, brokers, and other single-task real estate professionals.

The Bottom Line

In recent years corporate America has been subjected to repeated criticisms concerning its competitive position in the world. Shaken from complacency, companies in the United States have responded with bold, if not even draconian, measures to enhance their competitiveness. The adoption of new technologies has been and continues to be an especially successful strategy. Applying GIS tools to the management of a company's real estate assets can lead rapidly, and literally, to big dividends.

ADDITIONAL PRIVATE SECTOR APPLICATIONS

Data Vendors

Because real estate is a game of information arbitrage, a ready market exists for vendors of real estate data. Such vendors typically package and sell data on individual properties that are derived from public records and enhanced with proprietary collection and verification procedures. The information usually includes the following:

- Physical characteristics (e.g., land area, building size, year built, type of construction, parking spaces)
- Name and address of the owner
- The property tax assessor's valuation of the land and improvements
- Transaction details (e.g., when the property was last sold, sale price, seller, mortgage amount)
- The property's location as shown on a standard street map or on a property tax assessor's parcel map
- One or more photographs of each property's exterior and interior

Less frequently, a vendor will specialize in:

- Current and pending construction, as discovered from various permits filed with government agencies, interviews with local architects and construction contractors, etc.

- Historic, current, and projected pro forma information—e.g., asking rents, vacancy rates, operating expenses—at the MSA, submarket, or property level (see Figure 2.5)
- Individual leases, such as the floor-by-floor leases in place for every Class A office building in the central business district of a given city

These databases are always computerized. Clients can purchase the data in hard copy form by mail or fax. Frequently, but not always, they also can purchase and download the data electronically, on a turnkey medium (disk, tape, CD-ROM) or over the telephone line by modem.

Most of the nation's major data vendors already have a GIS or are actively investigating the technology for the following reasons:

- The vendor's databases are already computerized and location specific, so they are highly compatible with GIS.
- GIS makes data more attractive and useful to clients (again, sizzle and steak) because the clients can see the distribution of the properties on a map; compare the information to data in other GIS layers such as demographic trends, traffic volumes, and flood zones; and print superior presentation graphics.
- The vendors can expand their businesses by offering supplemental GIS data layers (e.g., flood zones) and adjunct services such as preparing hard copy GIS maps for clients who do not have their own GIS.
- They must stay abreast of competitors who offer GIS-related products and services.

The leading vendors of commercial real estate supply and demand data based on primary research at the property level include Torto Wheaton (Boston), REIS Reports (New York), and COMPS, Inc. (San Diego). The principal repackagers of governmental property records nationwide are DataQuick (San Diego), Transamerica Intellitech (Sacramento), Experian (Boston), and Lexis-Nexis (Dayton, Ohio).

The number and diversity of GIS products and services is increasing rapidly. This trend will continue to the benefit of all real estate professionals.

Mortgage Underwriters

The most far-reaching use of GIS technology in real estate to date has been its use by financial institutions to comply with the Community Reinvestment Act (CRA) and the Home Mortgage Disclosure Act (HMDA). GIS technology includes remarkably efficient and credible tools for identifying deliberate or de facto redlining practices and for comparing a financial institution's geographic sources of assets to its geographic distribution of products and services.

On a more sporadic basis, GIS technology has been used in facilities management in locating new branches, target marketing, mergers and acquisitions, risk assessment of mortgage portfolios, and other activities.

Moving forward, the most significant GIS application may well be GIS-based automated appraisal, as discussed by Mike Robbins in Chapter 4.

As a sign of the growing significance of GIS in this industry, the Fannie Mae Office of Housing Research sponsored a one-day workshop in February 1996 on "Research and Business Uses of Geographic Information Systems in Housing and Mortgage Finance." The well-attended workshop featured more than 25 speakers from the private, public, and academic sectors. Table 2.1, which delineates several dozen GIS applications in the mortgage finance industry, is reprinted from the introductory paper presented at the workshop.

Figure 2.5

Vacancy Rates in Downtown Chicago, February 1995

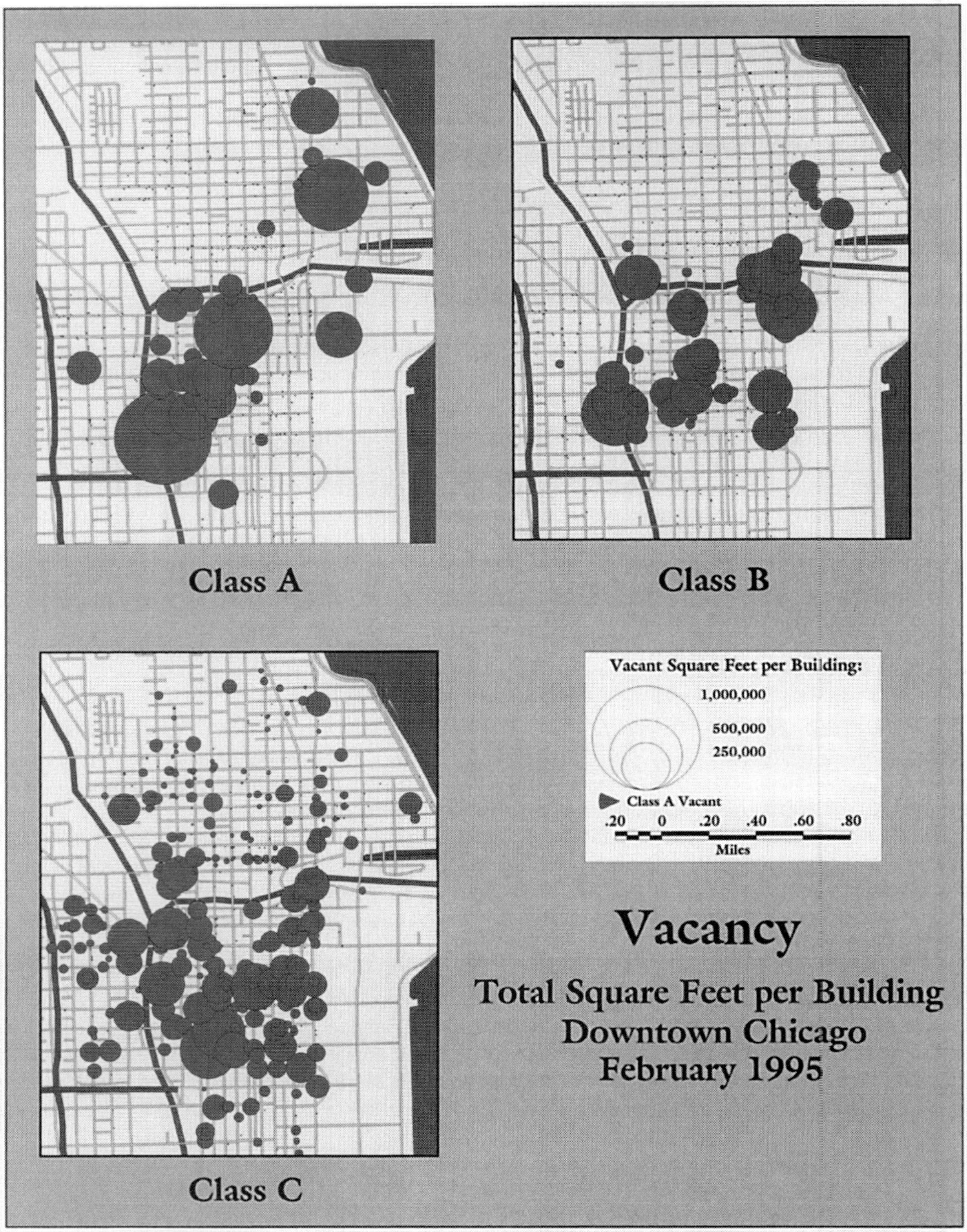

Table 2.1

GIS and Mortgage Finance Business Functions

Function	Who?	GIS Application
Risk Management:		
• Underwriting	• Mortgage lenders, originators, insurers and secondary market • Property insurers	• Automated appraisals using sales comparison or repeat sales approach • Automated address-based underwriting system
• Credit risk pricing	• Mortgage lenders, originators, insurers and secondary market • Property insurers	• Local credit risk modeling • Automated address-based pricing system
• Loan loss mitigation	• Mortgage lenders, servicers, insurers and secondary market	• Automated default tracking and mapping systems • Automated early warning, local default systems based on automated appraisals
• Loan loss reserves	• Mortgage lenders, originators, insurers and secondary market	• Automated appraisals to mark portfolio loans to market • Local credit risk modeling
• Prepayment risk pricing	• Mortgage lenders, servicers, insurers and secondary market	• Local prepayment risk modeling
• Agency risk/counter-party quality control	• Mortgage lenders, originators, insurers and secondary market • Property insurers	• Local credit risk modeling • Automated appraisals using sales comparison or repeat sales approach • Automated default tracking and mapping systems • Automated address-based underwriting systems
Marketing:		
• Market analysis	• Mortgage lenders, originators and real estate brokers • Mortgage lenders, originators, and secondary market	• Automated market trade area delineation and mapping • Automated area market potential estimation and mapping by mortgage type • Automated estimation and mapping of area potential mortgage and market demand actually served • Automated area market share analysis • Automated new market area assessment • Automated identification of underserved mortgage markets • Causal modeling of underserved mortgage markets • Causal modeling of local variations in mortgage market share
• Target and direct marketing	• Mortgage lenders, originators, and secondary market • Real estate brokers	• Area, customer, and business-based market segmentation using MCIF • Automated direct mail marketing • GIS-based multiple listing services with spatial display and analysis
• Delivery network planning	• Mortgage lender, originators, real estate agents and property insurers	• Location-allocation modeling • Identification of branch area trade area overlaps

Table 2.1

GIS and Mortgage Finance Business Functions (continued)

Function	Who?	GIS Application
• Market strategy	• Mortgage lenders, originators, insurers and secondary market	• Market analysis applications • Delivery network planning for mergers and acquisitions • Automated local market potential and share projections • Analysis of spatial variations in the price elasticity of demand for mortgage products • Simulation-based marketing spatial decision support systems
Regulatory Compliance:		
• Community Reinvestment Act	• Depository institutions	• Spatial display and analysis of lending, service, and investment patterns relative to low- and moderate-income neighborhoods
• Capital reserve requirements	• Financial institutions with whole mortgage loans in portfolio	• Automated appraisal systems for marketing mortgage loans to market value
• Underserved area goals	• Government sponsored enterprises	• Automated system to track and project loan purchases by areas defined as underserved by regulations
Research and Development:		
• Housing market research	• Large mortgage lenders, insurers, originators, and secondary market purchasers	• Full GIS spatial analysis capabilities
• Mortgage market research • Capital market research • Application development		

Source: Eric S. Belsky, Ayse Can, Isaac F. Megbolugbe, "A Primer on GIS in Mortgage Finance;" presented at Fannie Mae Research Roundtable Series workshop on "Research and Business Uses of Geographic Information Systems in Housing and Mortgage Finance," Washington, D.C., February 15, 1996

Developers

In general, developers have not been extensive GIS users for a number of reasons:

- Development activity in the United States has slowed in the past decade due to a glut of commercial space in most markets.
- Traditionally developers build whenever and wherever financing can be acquired. They do not always rigorously analyze current and projected real estate supply and demand.
- Most developers are oriented toward computer-aided design (CAD) systems rather than toward GIS for preparing site plans and blueprints.

A few development companies did embrace GIS early on. Homart Development Corporation was one of the first to use GIS for determining where to construct or rehabilitate major shopping centers. The Prime Group, Sam Zell, and Trammell Crow are other well-known developers with GIS capabilities.

Construction cranes are once again appearing in U.S. cities and financial institutions are becoming more GIS literate, so it is likely that more and more developers will adopt GIS technology. Moreover, trends suggest that real estate is increasingly evolving into an industry driven "by the numbers," rather than "gut feel."

Consultants

A dozen or so consulting firms in the U.S. specialize in helping real estate organizations evaluate and implement GIS systems, integrate GIS into their existing computer systems, and program customized GIS applications. Several of the authors of this book have consulting firms; others include Spatial Insights (Vienna, Virginia), Integrated Technologies (Newport Beach, California), Castillo Company (Phoenix), Geonomics (Boston), and Daniel Consulting Group (Austin, Texas). Most of the real estate consulting divisions of big six accounting firms and other management consulting companies have real estate-related GIS capabilities.

A smaller number of firms focus on providing sophisticated, GIS-based site selection services to major retail chains such as Circuit City, Office Depot, and Barnes & Noble. Leaders in this field include Staubach Retail Services (Dallas), Thompson Associates (Ann Arbor, Michigan), Terranomics, and Farallon Real Estate Services (both in San Francisco).

The major vendors of GIS software offer consulting and programming services to their clients, as do value-added resellers (VARs) and freelance programmers. Although they are knowledgeable about GIS software, hardware, and data, these consultants and programmers rarely have real estate-specific expertise.

As the number of real estate organizations using GIS increases, the number of consultants may also increase. On the other hand, if GIS tools become subsumed in special-purpose applications, especially "apps" designed for the Internet and intranet, fewer and fewer companies will need complicated, generic GIS installations.

Timber Companies

Few people realize that, among private companies, timber firms have the largest land holdings in the United States. Consequently, firms such as International Paper, Boise Cascade, Mead Paper, and Westvaco adopted GIS early, in the 1970s.

GIS technology helps timber companies keep track of their holdings by automating their mapping departments. Of even greater benefit, however, are GIS tools that help timber companies

- Optimize harvest schedules. GIS can link linear programming models to timber stand maps.
- Combat insect infestations and other natural threats to the timber inventory.
- Swap parcels with other timber companies and with the U.S. Forest Service.
- Comply with environmental restrictions, such as those that protect endangered species habitats.
- Identify opportunities for alternative, higher uses of land such as recreational home development.

PUBLIC SECTOR APPLICATIONS

The largest number of GIS users have been and will continue to be government agencies responsible for urban planning, public works, environmental protection, and parks and recreation. These agencies are a key part of the real estate industry since they influence (and sometimes actually decide) when and where real estate development will occur. Innumerable books, periodicals, conference proceedings, and other publications address GIS applications by these agencies.

Property tax assessors are the public sector officials whose activities most closely parallel real estate professionals in the private sector. The previous section on real property valuation discussed GIS applications relevant to assessment, including computer-assisted mass appraisal (CAMA). Property tax assessors use GIS not only for valuing real property but also for maintaining and updating their parcel maps. Indeed, many systems are budgeted as automated drafting devices and used secondarily as analysis tools.

Less numerous but no less intriguing public sector applications for GIS can be found in economic development, asset management, and data sales.

Asset Management

Asset management in the public sector has three principal components. The first is the management of natural resources without commercial value, including wilderness preserves, parks, weapons testing areas, and so on. A GIS can certainly be used to inventory and manage these assets, but such applications are not the focus of this book.

The second component is the management of natural resources with commercial value, e.g., lands suitable for grazing, farming, or similar purposes. At the federal level, agencies such as the Bureau of Land Management and the Department of Agriculture extensively employ GIS to manage these resources. At the state level, various state agencies in Massachusetts, Texas, and Hawaii were pioneers in utilizing GIS for inventorying, valuing, managing, leasing, swapping, and selling state-owned properties. Few agencies at the local level administer natural resources, with or without a GIS.

The third component of public sector asset management relates directly to the private real estate industry and consists of office buildings, warehouses, and other commercial properties owned or leased by government agencies. Government workers account for approximately 20% of the nation's work force, so the amount of space needed by federal, state, and local agencies is substantial. Inefficient use of space can cost taxpayers a considerable amount of money. GIS technology can be used by government agencies to combat inefficiencies just as it is used to increase productivity in the corporate real estate market. As of this writing, the General Services Administration, which manages, leases, and sells buildings belonging to the U.S. Government, is investigating linkages between GIS and property management software packages.

Economic Development

The mission of an economic development agency is to convince companies to locate or expand their business operations in the agency's service area, which is usually a state, county, or city. Such private investments enhance the area's employment base, household incomes, property values, and tax revenues. To accomplish its mission, the agency must have ready access to diverse and extensive data of interest to prospective companies, including information on

- Labor force size, composition, training, and availability; union memberships; and average wages
- Utility rates for electricity, gas, water, and sewer service
- Tax rates on real property, equipment, and payroll
- Transportation access by highway, air, railroad, and sometimes waterways

To be successful, an economic development agency must be able to present information in a compelling way, to distinguish itself from the innumerable other economic development agencies that are recruiting the same companies. Once again, the ability of GIS technology to provide "the sizzle as well as the steak" can prove very beneficial.

At present, the most advanced user of GIS for economic development is probably the State of South Carolina, which was able to attract the first BMW factory outside of Germany. The content and structure of South Carolina's GIS databases are similar to the in-house GIS data- bases of major retailers, brokers, and other organizations involved in site location decisions. (See Figure 2.6.)

Viewing a map of the entire state on the computer screen, for example, a client might first ask for all counties that have average wage rates below a certain amount, an unemployment rate above a certain amount, no membership in one or more specified labor unions, at least one interstate highway interchange, and one commercial airport. Counties fulfilling the criteria are highlighted automatically on the screen. The client can then modify the criteria and immediately display the results. The process continues until the client has identified the most promising counties.

Next, the GIS zooms into each of the counties selected, one at a time, to display more geographically finite information. The client might look at labor force characteristics at the city, zip code, census tract, or even block group level; the boundaries of various tax assessment districts; the actual locations of available buildings; the locations of all nearby streets, railroads, water and sewer lines; and the electric grid. The client can use these map displays to find specific, suitable sites in each of the priority counties.

Finally, the client can retrieve data about each suitable site, usually by clicking on the site on the computer screen. Information about building attributes such as asking rent, the amount of

Figure 2.6
South Carolina Infrastructure/Economic Development Program

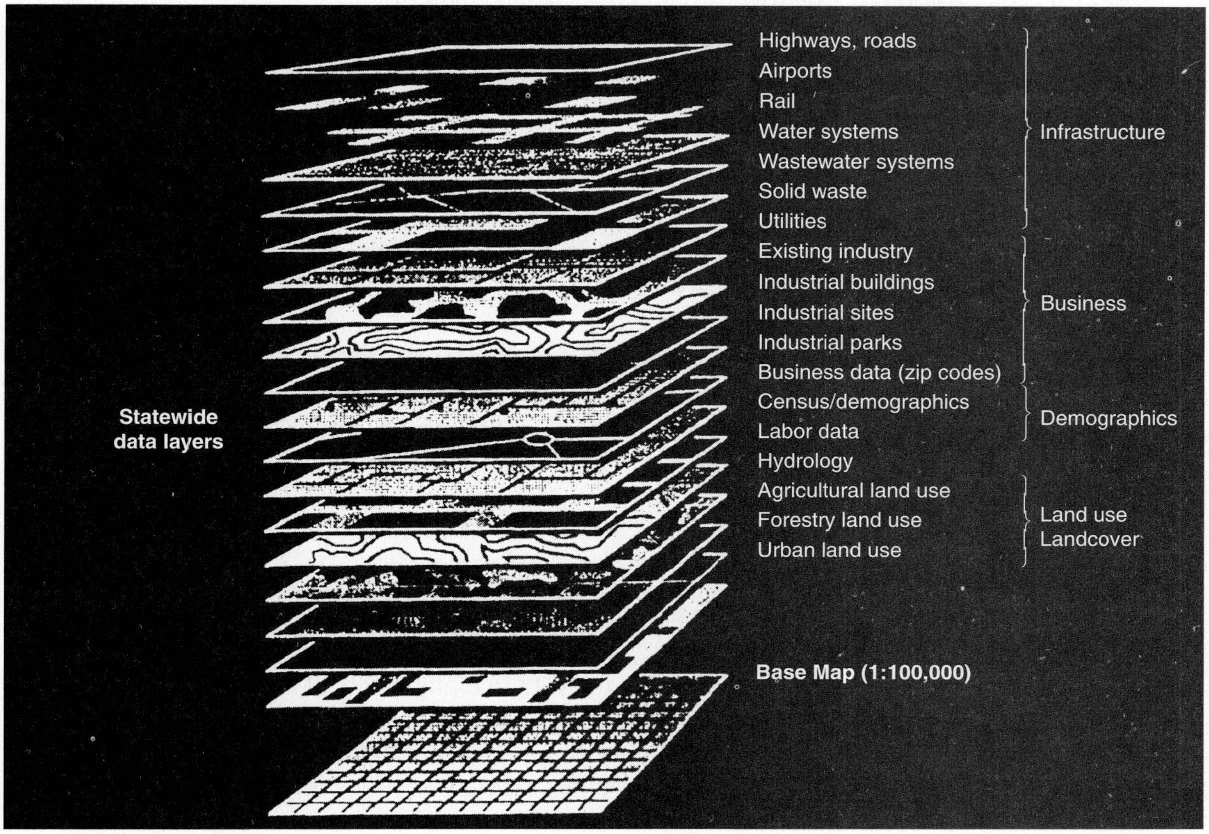

contiguous space, the number of loading docks, ceiling heights, and the year and type of construction are instantly available. In some cases the client can even view and manipulate a CAD display of the floor plan and surrounding infrastructure to determine how walls could be rearranged to accommodate the client's business operations. (Springfield, Oregon, did this in recruiting a Sony optical disc manufacturing facility.)

By using this GIS procedure for finding a suitable facility instead of traditional methods, the economic development agency and the prospective client can save an immense amount of time. Moreover, the GIS methodology is much more credible than a traditional approach involving paper maps and filing cabinets. The competitive edge enjoyed by economic development agencies with a GIS is obvious.

Government Agencies Selling GIS Data

A tremendous number of maps are encoded into a GIS by state, county, and city agencies, and the number is growing every year. Such maps may show water and sewer lines, current zoning, existing land use, topography, environmental hazards (e.g., flood zones, earthquake faults, toxic waste sites), traffic volumes, crime locations, and other features. Selling these GIS databases in hard copy and/or electronic form to private sector organizations and individuals has become a very hot topic in recent years for the following reasons:

- Public officials have had to scramble for additional sources of revenues in tight budgetary times, so potential revenues from user fees are a politically attractive option.
- Agency investments in GIS databases have reached hundreds of thousands, millions, and even tens of millions of dollars. Finding ways to amortize these investments has become a priority.

- Brokers, bankers, appraisers, real estate attorneys, title companies, civil engineering firms, and many other private sector organizations have expressed serious interest in purchasing digital and attribute data (e.g., computerized maps of all assessors' parcels and associated tabular data on each parcel's ownership, size, valuation, etc.).

Although sales of GIS data could be a major "win-win" for both public sector sellers and private sector buyers, only a handful of agencies across the nation have established such operations. The obstacles are considerable:

- The federal government's policy is to view data as a "public good" that should be distributed for the cost of reproduction only; many states have similar policies.
- Government agencies are typically not set up to operate as for-profit entities. They are not usually staffed by people skilled or interested in entrepreneurship and, in any event, should not be competing with private industry.
- Daunting legal issues arise regarding warranties and liabilities, confidentiality and the right to privacy, freedom of information, copyrights, and trade secrets, etc.

As more agencies find ways to overcome these obstacles, an ever-greater number of real estate professionals will benefit from convenient, cost-effective access to extensive GIS databases.

For more information on GIS in the public sector, readers should contact the Urban and Regional Information Systems Association (URISA) in Park Ridge, Illinois, which is the oldest (35 years old) and largest (3,000 members) professional society of its type. The association's secretariat can be contacted at 847-824-6300 or via the Internet at www.urisa.org.

CONCLUSION

This chapter has clearly demonstrated that GIS is relevant to many facets of the real estate industry. The most popular GIS functions, such as thematic mapping, have existed for more than a quarter of a century. Relatively inexpensive software and data on personal computers have been available for nearly a decade. Nonetheless, real estate professionals have been very slow to adopt the technology. Those who have, mainly brokers, often have focused their attention on presentation capabilities rather than the technology's more powerful data integration and analysis functions (i.e., choosing the sizzle over the steak).

Why have real estate professionals been slow to pick up on GIS? Partly because any new technology takes time to catch on. Alexander Graham Bell had tremendous difficulty finding investors for his telephone invention. Hardly anyone envisioned the usefulness of such technology when the mails worked just find. Color television existed for 60 years before a significant number of households traded in their black and white sets. The facsimile machine was around for 30 years before businesses and households began to appreciate its usefulness.

Even when the potential benefits of a new technology are apparent, few business men and women in any industry are comfortable being pioneers. Instead, they let others take the risks and embrace a new technology only when the competition forces them to do so.

GIS in real estate is just reaching that critical stage. It is such a powerful technology—for both presentation and analysis purposes—that as soon as one major player adopts it everyone has to pay attention. Given a technology adoption curve that is geometric, the industry now appears to be leaving the level plain and starting to climb the hill. The choice confronting commercial real estate professionals used to be: "Do I want to be a GIS leader or a follower?" The choice now is: "How much longer can I remain competitive without GIS?"

chapter 3

Residential Real Estate and GIS

by Marcie Wallis and Anthony Hutchinson

Real estate and GIS look like a very good match once one begins to explore the benefits of their union. The commercial real estate industry has used GIS in the site selection process for a few years now, so why does the residential real estate industry hesitate? Why don't residential brokers embrace this technology and use GIS data in the home selling and home buying processes?

Unfortunately, the residential real estate industry has been slow to recognize or understand the benefits of GIS, and getting GIS adopted is a very tough prospect. Many residential Realtors® already have a proven technique for home buying and selling; they do not believe they need to improve to be successful. Moreover, these same Realtors®, whose median income was $33,500 in 1995, do not have time to experiment with an unproven business tool. The bottom line for residential Realtors® is to "list" a home that will eventually lead to a sale.

Despite these obstacles, there is a growing awareness of GIS technology in the business community. This awareness is beginning to trickle down into specialty business fields, especially real estate. Experienced brokers, brand-new real estate professionals, and real estate technology/information innovators are beginning to consider the benefits of using GIS as a business tool for residential real estate. The National Association of Realtors®, the major trade association for the industry, has recognized the benefits of GIS and is in the process of exploring these benefits as they pertain to the residential real estate industry.

THE NATIONAL ASSOCIATION OF REALTORS®

Residential real estate is a very diverse industry, made up primarily of people who have chosen real estate as a second career. In 1995 only 6% of the Realtors® in the industry were embarking on their first career. The remainder moved into real estate as a second or subsequent career, and their motivation and work ethic are different. Understanding these professionals is vital to promoting the use of GIS to residential Realtors® and to motivate them to be GIS real estate application pioneers.

The National Association of Realtors® (NAR) is the nation's largest professional trade association, with nearly 700,000 members, most of whom are involved in residential real estate. The Association was founded to provide knowledge, representation, association, and business services to its membership.

The NAR membership is organized into 1,800 local boards of Realtors® and state associations of Realtors®. Every NAR member (Realtor®) belongs to a local board and a state association. Under the NAR umbrella, there are seven affiliated societies, institutes, and councils that cover the major specialty areas of the real estate industry:

1. Counselors of Real Estate (CRE)
2. Commercial Investment Real Estate Institute (CIREI)
3. Institute of Real Estate Management (IREM)
4. Realtors® Land Institute (RLI)
5. Realtors® National Marketing Institute (RNMI)
6. Society of Industrial and Office Realtors® (SIOR)
7. Women's Council of Realtors® (WCR)

In addition to overseeing public policy issues, NAR is responsible for keeping its membership informed of changes in business procedure and advances in technology. NAR also helps its members realize the power they control in the form of information—especially property information. Anyone who works in real estate knows about the MLS (multiple listing system) that Realtors® use to access vast amounts of property information. This information is a natural link between the real estate industry and GIS. Linking GIS and MLS data is the first logical step in incorporating GIS into residential real estate.

GIS PRODUCTS FOR THE REALTOR®

At its conception, the multiple listing service could be characterized as proprietary, with listing books, black and white photographs, car tours of neighborhoods, and local courthouse records. Today MLS is computerized, public, virtual, realistic, global, on-line, and geographic information-oriented. The nature of this new MLS marks a fundamental shift in the thinking of Realtors®.

Moving MLS technology forward moves Realtors® forward because the MLS is their primary tool in the home buying and selling process. With this is mind, it stands to reason that one of the first and most common uses of GIS in residential real estate is as a link to the MLS.

A number of GIS vendors have developed this link and are marketing the capability to local associations of Realtors® and regional multiple listing services as well as some large franchises, as described below.

Vendor-Developed Systems

Northwest GeoGrafx

Developed in cooperation with the Puget Sound Multiple Listing Association, Northwest GeoGrafx has developed its Latitude program, which links information on properties with maps to help real estate agents sell homes. Using a laptop computer, agents can type in the features their clients are looking for in a new home and display detailed maps and textual information (e.g., list price, address, number of bedrooms and bathrooms, age) to pinpoint the homes that most closely match the description. Agents can tailor the maps to show a broad or specific area, identify and prioritize homes of interest using a variety of symbols, and overlay some of the textual information onto the map.

Latitude was first tested in Seattle in 1992, and it has become so popular since then that it is now being used by more than 60% of all real estate offices in the area. The program is beginning to be used by agents in Atlanta, Chicago, and other metropolitan areas as well. Future developments by Northwest Geografx will include color photographs of each home in the program and possibly on-line video tours of the homes and their neighborhoods.

REMAP Corporation

The REMAP Corporation was established in 1991 to provide state-of-the-art marketing tools to real estate professionals. The company provides custom software, data conversion services, hardware/software product marketing, and information to the real estate professional.

REMAP has developed two products for the residential market. The first is IRIS (Integrated REALTY Information System™) an on-line system that integrates MLS information into a GIS environment containing area maps, census demographics, tax record data, and photos. The system also allows the user to perform desktop publishing tasks and prepare custom reports for clients.

The system works in a Windows environment and will run on a computer with a 386 processor or better. Systems are in place in Mankato and Minneapolis/St. Paul, Minnesota, and in Brevard County and Tampa/St. Petersburg, Florida.

The second REMAP product is ADD-MAPS (Automated Digital Data Maps). This system is a stand-alone product that provides much of the functionality of IRIS but without the constant contact with the MLS.

This desktop publishing mapping system allows the user to input MLS data via a dBase file and integrate that information with built-in census, demographic, and other local data. In addition, the system has an automatic address-matching capability. As with IRIS, ADD-MAPS works in a Windows environment and requires a 386 machine or better.

Waypointe Technologies, Inc.

Waypointe Technologies, Inc. produces Navigate, a software application that will interface with the MLS and provide custom comparative market analysis (CMA) maps, custom touring and showing maps, MLS photos, and custom demographic information.

Navigate allows the real estate professional to locate and display properties in the MLS. Realtors® are then able to point and click on a picture of a home and retrieve vital MLS information. Users can choose up to eight homes, view each in a window, and print off a CMA in a format customized by the Realtor®. Waypointe has also integrated U.S. census information into the system so prospective buyers can get a better understanding of the areas where the homes are located. An address-matching feature allows Realtors® to locate the properties in question.

Proprietary Computer Systems

In addition to vendor-produced computer systems, some real estate firms are developing proprietary computer systems, with GIS as a prominent feature. Two of these companies are discussed below.

Burnet Realty and Edina Realty of Minnesota, two of the country's top 10 real estate brokerages, have developed their own GIS-related computer systems. Both companies came to the conclusion that computerized services would help them to operate more efficiently, attract more customers, and complete more transactions.

Burnet Realty

Burnet Realty was the first of the two firms to enter this revolutionary era, releasing their system on November 3, 1994. Burnet enlisted the help of 11 vendors and spent more than two and one-half years producing the Radix system. The Radix system downloads the home inventory and property photos from the regional multiple listing system and combines these data with other databases that contain information on neighborhoods; pricing; proximity to schools, parks, transportation, and industry; age of housing; and many other variables of interest to prospective home buyers.

The focal point of the system is its ability to display this information in a map format. Radix creates a map base and overlays the information agents use in sales presentations. More important than its ability to provide great presentation material, the system allows agents to query the MLS database on a number of different client criteria, display the homes on a map, and show a photo of the property.

Burnet agents can retrieve information describing each home's respective neighborhood and city, such as their population density, percentage of children, and property tax ranking. The system also features desktop publishing capabilities, mortgage scenario calculations, sample contracts, and an optional tie-in to the U.S. Global Positioning System, which allows the agent to monitor his or her movement in relation to properties of interest. This product is not available to the public.

Edina Realty

Edina Realty teamed up with U.S. West to develop a system that provides the home buyer, working with a real estate professional, easy access to detailed real estate information that, until recently, was not centrally located. On January 17, 1995, Edina Realty and U.S. West announced the creation of AgentEdge. This new product is a combination of GIS and MLS data with U.S. West's proprietary, 14-state customer files.

AgentEdge provides access to vital information on property features, taxes, demographic data, neighborhoods, and schools. The system can also provide personalized financing option reports, e.g., rent versus buy, fixed mortgage versus adjustable mortgage, tax analysis. Moreover, the system includes maps illustrating the location of parks, schools, public transportation, shopping districts, and places of worship in relation to homes accessed from the regional MLS. Another function allows the agent to calculate mileage to any feature of interest on the map.

An agent can access this system in-house on a PC or on a laptop computer via cellular phone while in the field. Presently, this system is available for real estate professional use only; however, both Edina Realty and U.S. West have plans to offer a version of this product to the public.

GIS TECHNOLOGY MISHAPS

For every residential real estate GIS application that has enjoyed success, there are a few that have not. The idea or application may be sound, but the users of the application may not be prepared for the technology. Two cases in point are described below.

HomeView Realty Centers, Inc.

HomeView is a company based in Needham, Massachusetts, which captured the real estate market's imagination in the early 1990s with its innovative, computerized style of home selling; by the autumn of 1994, however, it was facing financial difficulties. The company's evolution into a competitor of established brokerage firms is blamed for its problems. Ultimately, however, the technological advances the firm brought to the real estate industry—e.g., client criteria based on computer searches for the appropriate home, touch-screen kiosks in malls and airports, and color photo tours of selected homes—have been able to bring about the company's rebirth. The new focus of HomeView is to establish partnerships with large real estate firms that use its information system to sell homes.

Here is how the service works. In consultation with a HomeView representative, potential homebuyers rank their preferences for a new home according to desirability and these preferences are then entered into the sponsoring broker's database. More than 300 characteristics can be indicated, including the size of the home, architectural style, preferred towns, distance from transportation, schools, and the range of home prices. A touch-sensitive monitor then displays a map of the available homes fitting the buyers' specifications. Each home is presented with several color photographs of the building's interior and exterior. Other screens show detailed maps of the home's location within the town, county, or state as well as demographic information about the area. At the end of the session, the HomeView system prints a four-color booklet with photographs and information about each home the buyers would like to visit.

The system was built to provide the home-buying public with a quick way to view numerous homes and limit the number they actually visit. Business Wire, Inc. reports: "Surveys state that the average home buyer currently visits between 25 and 40 homes before selecting a home. Based on experience in the Boston market, home buyers utilizing HomeView visit between 6 and 12."

HomeView Realty Centers, Inc. is partnered with real estate firms in New York and Northern California. The revitalized company is expected to form partnerships in other major U.S. cities.

Real Estate Information Network (RIN)

In the mid-1990s, NAR launched the Real Estate Information Network (RIN) in conjunction with several vendors of GIS software and data—Environmental Systems Research Institute (ESRI), Geographic Data Technology (GDT), and Environmental Risk Information and Imaging Services (ERIIS). RIN was designed to provide the membership with a number of on-line services, including the ability to run simple queries on demographic data and to generate maps and reports on areas of interest throughout the United States. These services were tied into a national on-line MLS.

In the summer of 1996, NAR's Board of Directors held an emergency meeting to decide the future of RIN. After reportedly investing $18 million, with more millions needed, NAR questioned whether RIN would ever be financially viable. A key concern was the number of individual residential brokers who would invest hundreds or even thousands of dollars in the desktop software and hardware needed to connect to RIN. NAR decided to seek other investors.

Ultimately NAR sold RIN to RealSelect, retaining a 15% stake in the company, which now operates www.Realtor.com. RealSelect currently offers a small subset of the original RIN services. The bottom line for RIN was the price; no one questioned the usefulness of the services. GIS, like any new technology, must show an attractive return on investment (ROI) if it is to be widely adopted.

THE INTERNET

The Internet, and especially the World Wide Web, is dramatically changing the real estate industry in general and residential brokerage in particular. Multiple listing services and boards of Realtors® nationwide have a love-hate relationship with the Internet. They love the Internet because brokers can more easily remain current on market conditions, gain a wider audience for their listings, and save time by not driving potential homebuyers from one home to another. On the other hand, they hate losing their information monopoly and the fact that home buyers and sellers need fewer services from them.

Regardless of their feelings, residential real estate professionals know the Internet is here to stay, and virtually all brokers are contemplating placing their listings on the Web. Some of the largest, most visible Internet companies in the world, including Microsoft and Yahoo, are positioning themselves for a piece of the residential brokerage market. The Web sites created by these companies provide

- Listings that meet the criteria typed in by the potential buyer, i.e., a certain price range, the desired number of bedrooms and bathrooms, and amenities such as a swimming pool
- Neighborhood analyses, including demographic profiles, school statistics, crime reports, and environmental hazards
- Links to home buying services, notably mortgage companies, but also moving companies, insurers, local brokers, etc.

The following home search sites relating specifically to GIS had an early presence on the Internet:

- www.cyberhomes.com is the site with the most complete mapping capabilities. As shown in Figure 3.1, a cyberhomes map of the San Francisco Bay area taken directly from the Internet, this site allows users to search for homes for sale by clicking on maps on various geographic scales to zoom into an area of interest. The site also supports radius searches, e.g., all homes for sales within two miles of a user-designated location. All homes fulfilling a buyer's criteria are plotted on a street map, along with school locations and other important reference features.
- www.Realtor.com is the official site of NAR and the inheritor of RIN's functionality. The site uses GIS maps to zoom into an area of interest, but not all MLSs contributing to RIN allow display of a map showing the actual location of homes for sale. Figure 3.2 is the product of a www.Realtor.com search that shows street-level locations, in this case in the eastern part of

Figure 3.1

Cyberhomes Map of San Francisco

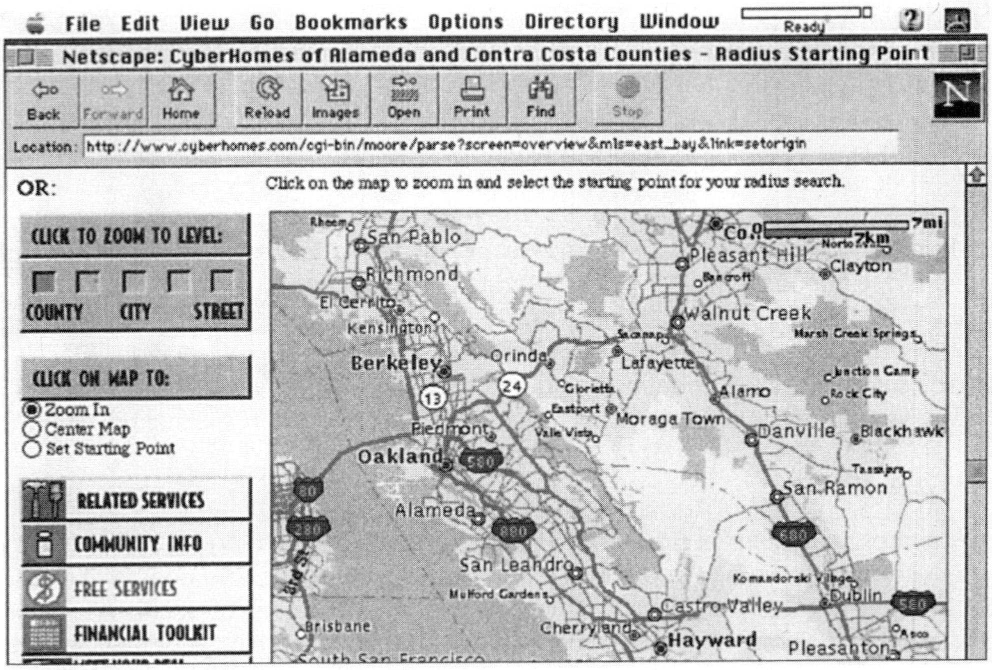

Figure 3.2

Map of Homes for Sale

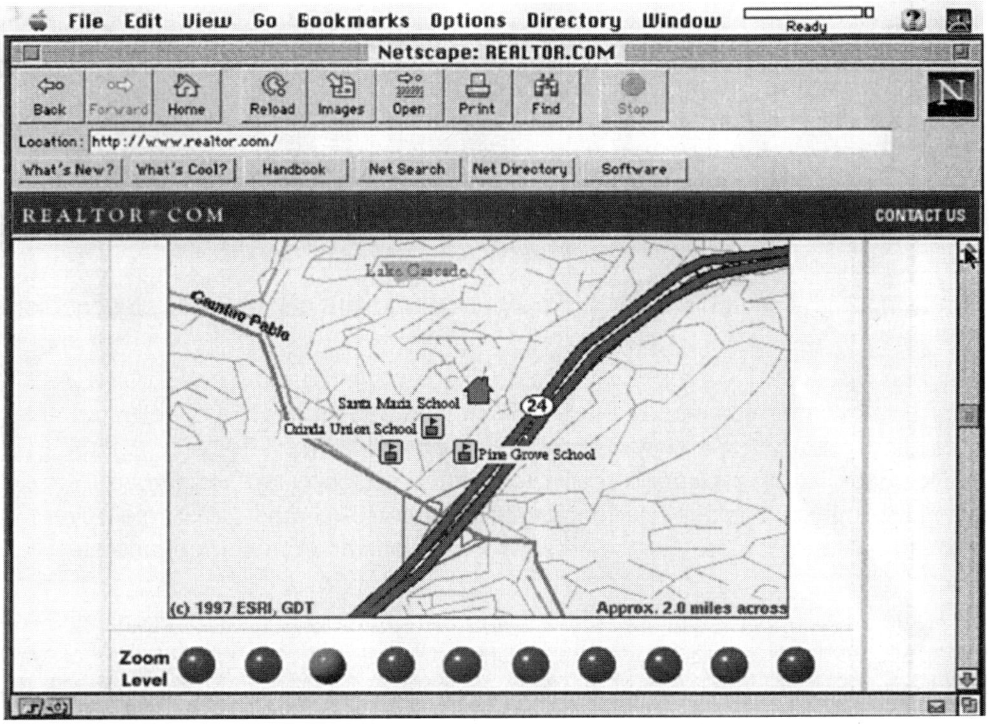

the San Francisco Bay area. Note that the site depicts not only the location of a home for sale that meets the buyer's criteria but also nearby schools and amenities. As of June 1998, the site had more than one million home listings nationwide and was growing rapidly.

- www.mapquest.com was one of the first Internet sites to offer mapping. Each year the site has millions of visitors, who can locate a specific address on a map and plot the most efficient route from one location to another. In August 1997, the site expanded into the residential property transactions arena with its "homes" section. House and apartment listings, mortgage and insurance information, community and school profiles, and other information are all available and frequently can be mapped at this site.

THE GIS FUTURE FOR RESIDENTIAL REALTORS®

Residential Realtors® may or may not embrace GIS technology. The make-up of the industry is diverse, and those who are continually looking for new and efficient ways of buying and selling homes will use the technology. Those brokers who are happy with the way they are presently buying and selling homes will continue as usual.

The best opportunity for GIS is to make its way into the residential real estate industry right now—through the Internet. Many Realtors®, whether they are new, old high tech, or low tech, are interested in the business opportunities the World Wide Web provides. Companies like GeoSystems (Mapquest Web site) and Lysias (Know Thy Neighborhood) stand to gain the most Realtor® business. Hands-on technological education for real estate professionals and simplicity are the keys.

The greatest obstacle to incorporating GIS applications into the residential real estate industry is the complexity of the products. Realtors® do not have time to learn difficult software packages. Moreover, most real estate agents do not make enough money to hire a GIS specialist to produce data from a GIS product. To penetrate the real estate industry successfully, GIS corporate leaders and innovators must study and understand the industry and create products that meet its needs.

With those caveats in mind, some potential GIS applications for residential real estate are described below. Each promises a "bottom line" benefit—a necessity if Realtors® are to accept and use GIS.

Developing Market Territory

GIS allows residential Realtors® to identify the location of areas where they can successfully target markets. A broker can consider demographic trends and a new agent's lifestyle profile and then identify those market areas that would be most productive and suitable to the new agent. This would serve as a good starting point for a real estate professional just getting into the business. The idea of profiling the agent to fit a particular market area can also be reversed, with the agent targeting areas in which to generate business. For example, a broker can identify areas with market potential by studying the percentage of renters, the amount of rent paid monthly, years of home occupancy, and the age of homes.

Real estate professionals also can monitor areas of high population growth, which may suggest that they concentrate marketing efforts in those areas. The application described below, locating home buyers, is similar to developing a market territory, but this effort is focused on marketing one property.

Locating Potential Home Buyers

A Realtor® may single out a home that he or she has been entrusted to sell and build a profile of potential home buyers based on the people selling homes and the neighborhood. GIS will allow the Realtor® to query the general area, based on the demographic and lifestyle data at the zip code or census tract/block level, and locate pockets of people who may be ready to trade up or buy a first home. Once the pockets of potential buyers are located, a Realtor® can market the specific home to a narrower market, not the entire city or MLS region. This will reduce advertising cost (direct mail, flyers, etc.) while increasing the potential return on investment.

Locating Homes for Buyers

This is the application that Realtors® will find most useful. Most Realtors® do not prospect for buyers; rather, they prospect for home sellers. Residential brokers look for people who want to sell a home so they can not only list the old home on the market but also find them a new home.

Many Realtors® believe the market will take care of the "for sale" home, so they push to locate a new home for the client. The application described above demonstrates how one can move the old house using GIS; this example will show how quickly one can find a new home for a buyer.

The Realtor® can use the MLS to query the style and price of the home a client is interested in. This query will indicate a number of homes over the MLS area. GIS technology allows geocoding, i.e., providing a latitude and longitude coordinate for each home found in the initial query. Now the Realtor® can see the location of each home in relation to all others and, more importantly, in relation to secondary attributes that may be of interest to the client. For instance, how close is this home to public transportation? What schools are nearby? And is a shopping district within 10 minutes of the new location? This information will allow the Realtor® to reduce the number of homes the client will want to visit. The Realtor® could go one step further and, based on the client's lifestyle and basic demographic profile, determine which areas have homes that the client is most likely to purchase.

Optimal Routing of "Tours/Caravans"

Once the number of homes to be presented to the client is determined by the broker, he or she can use routing software within the GIS to determine the most efficient route and time schedule for showing the homes. This will help optimize the broker's time so he or she can see more clients and show more homes. Similarly, computer-generated road maps included in a potential home profile are used by some Realtors® to familiarize out-of-town buyers with the local area.

Listing Presentations

This GIS application allows the Realtor® to provide a client with a professional-looking and informative report on potential properties. Moreover, the agent has the ability to present findings to a client "on the fly." The real estate professional can run queries against the MLS database (e.g., to find comparable home sales prices) and present the information to a client right off the computer. This can save the Realtor® hours of research and free up time for sales opportunities.

Realtor® Association Applications

In addition to real estate professionals, Realtor® association staff can use geographic information technology to promote and protect NAR's membership. NAR lobbyists, researchers, marketers, economists, lawyers, state and local board presidents, and a host of others will find GIS useful in many of their day-to-day activities.

Realtor® staff can use GIS to quickly identify patterns in home pricing or demographics within their board market. This information will provide the state or local association leadership with insight into the real estate economy of their board jurisdiction at a number of levels. (Since boards usually encompass a county or two, demographic and socioeconomic analysis can be performed at the zip code, census tract, or census block group level.) Such studies allow the organization staff to create information products that provide the membership with a real estate market advantage. This application is beneficial not only to members within the state association or local board for which it is performed but also to companies that deal in the home buyer relocation market.

Realtor® state associations and boards can use GIS to monitor membership population trends. This function allows the staff to determine if offices and meetings are in locations that are accessible to a majority of the membership.

In addition to understanding the locational distribution of their members, boards can use GIS to analyze demographic and business characteristics for targeting and tailoring board products, programs, and services. For example, if a board sees that 90% of its members are interested in commercial activities and are concentrated in the western portion of the board's service area, staff may consider locating commercial educational courses and products at a site convenient to these members.

The previous examples represent a small portion of potential applications of GIS in residential real estate. GIS professionals are still in a period of discovery, finding out what GIS can provide. As the knowledge of Realtors® continues to grow, the industry will provide criteria to design and implement increasingly relevant GIS applications.

CONCLUSION

As with any endeavor that moves into uncharted territory, there is some resistance. However, society has indicated the need for technological change in every area, including real estate. The previous sections have shed some light on applications that are available for the industry today.

GIS offers something for everyone in real estate. More to the point, these applications are readily available and will benefit the residential real estate industry. Once embraced by industry professionals, GIS and related technology will be a big part of the new technological order within real estate.

chapter 4

Overview and Case Studies in GIS-Based Appraisal

by Michael L. Robbins, PhD, CRE

A geographic information system (GIS) is a set of computerized tools (including both hardware and software) for collecting, storing, retrieving, transforming, and displaying spatial data. Data in a GIS usually represent a model of the real world, describing real-life objects or geographic entities in terms of their position with respect to a known coordinate system on the earth's surface; attributes unrelated to position; and spatial mathematical interrelationships with each other through the concept of topology.

The utility of a real estate parcel and its value are directly affected by what is happening in the vicinity of the parcel. For this reason, real estate is recognized as a *spatial* commodity. The importance attached to "location, location, location" attests to real estate's spatial dependency. GIS technology provides real estate professionals with tools to help them study the spatial context of their decisions.

The use of a GIS can greatly decrease the time required to perform many types of conventional real estate analysis. GIS can improve the accuracy of analysis by allowing the calculation of measurements (distance, area, travel time, etc.). More importantly, a GIS can also enhance the interpretation of spatial data by facilitating analyses that would normally be difficult or impossible to perform (percent slope, relative elevation, market capture decay, etc.). With the improved productivity that a GIS offers, real estate professionals can undertake analyses that they previously would not have attempted, allowing them greater freedom to test new ideas and contribute to the evolution of innovative techniques.

To help differentiate among the complex array of applications to which GIS techniques can be applied in real estate analysis and appraisal, this chapter is divided into three sections. The first section introduces a general structure to various real estate analyses and the major participants in each. The types of data necessary to support the various analyses are explored.

The second section will demonstrate various types of analytical applications and associated data types with two case studies. The first case study concerns single-family housing within an urban setting—i.e., improved property analysis. This example will demonstrate the combining of TIGER/line data, assessor inventory information, commercial cost estimating software, and commercial socioeconomic profile data using the SPANS Map desktop mapping program.

The second case study will demonstrate the various types of analytical applications and associated data types applied to rural land analysis—i.e., unimproved property analysis. This example will utilize the TIGER/line data and the United States Geological Survey (USGS) digital elevation model (DEM) data. The case study will demonstrate terrain analysis capabilities using the MICRODEM software package.

The third section will explore a particularly exciting, but controversial, subject—automated appraisal, including the use of GIS tools to produce more accurate value estimates.

THE NATURE OF THE REAL ESTATE DECISION PROCESS

The material in this section has been abstracted and updated by the author from various lecture materials and class handouts distributed by the late James A. Graaskamp, former chairman of the Real Estate and Urban Land Economics Department at the University of Wisconsin-Madison Graduate School of Business.[1] Jim Graaskamp always began teaching real estate to new students with the following story:

> "Someone rolled a rock across the entrance of a cave and created an enclosed space for the family—a warmer, more defensible shelter, distinct from the surrounding environment. This can be called the first real estate development. Since then real estate activity has evolved and taken many forms to meet the needs of man and society. Once based on need and custom, real estate is now based on social economics and stature."

This statement captures the essence of modern real estate. Today, real estate analysis must take into consideration a broad spectrum of regulatory, market, and investor controls and objectives.

What Is Real Estate?

For the GIS enthusiast, it is interesting to note that real estate can be defined generally as space delineated by man, relative to a fixed geography on the earth's surface, and intended to contain an activity for a specific period of time. Note that, to the three general dimensions of space (length, width, and height), real estate has a fourth dimension—time of possession and benefit. This extra dimension gives rise to the space-time characteristic of real estate, which is illustrated by the use of such terms as square footage per year, apartments per month, motel rooms per night, and tennis courts per hour.

One fundamental attribute of real estate is that any space-time unit has a corresponding monetary value. While many of the value judgments and debates about real estate projects relate to elusive criteria of what is good and beautiful, in a money economy the ultimate criterion is cash. Thus the focus of real estate decision making is the conversion between space-time and money-time.

The complexity of the various interactions associated with this conversion is illustrated in Figure 4.1, which depicts the real estate process. This figure is a graphic representation of the three principal analytical focus areas of real estate decision making and the secondary socioeconomic systems affecting each. As the figure shows, the successful determination of a use for a particular site requires successful arbitration between the site user(s), the infrastructure serving the site, and the skills and capital necessary to produce and operate the improvements for the use.

Land Versus Location

The well-informed real estate analyst generally understands that land is not location, property rights, or the most important aspect of a project site; instead, land is a natural resource upon which real estate decisions come to bear. Land is brought under control to bear structures. It is a finite resource that can be exhausted by extractive industries, destroyed by seismic and ocean upheaval, or wasted by ignorance of its processes. It is a limiting factor in development and both a reference and a bearing point for space-time units.

A site is created by the land's physical attributes together with legal/political constraints, linkage attributes that define location, dynamic attributes that exist in the eye of the beholder (e.g., prestige, crime, view quality), and attributes of a larger environmental system.

1. Readers interested in additional writings of Graaskamp are directed to Stephen P. Jarchow, *Graaskamp On Real Estate* (Washington, D.C.: Urban Land Institute, 1991) and James R. DeLisle and J. Sa-Aadu, eds. *Appraisal, Market Analysis, and Public Policy In Real Estate: Essays in Honor of James A. Graaskamp* (The American Real Estate Society, Kluwer Academic Publishers, 1994).

Figure 4.1
The Real Estate Process

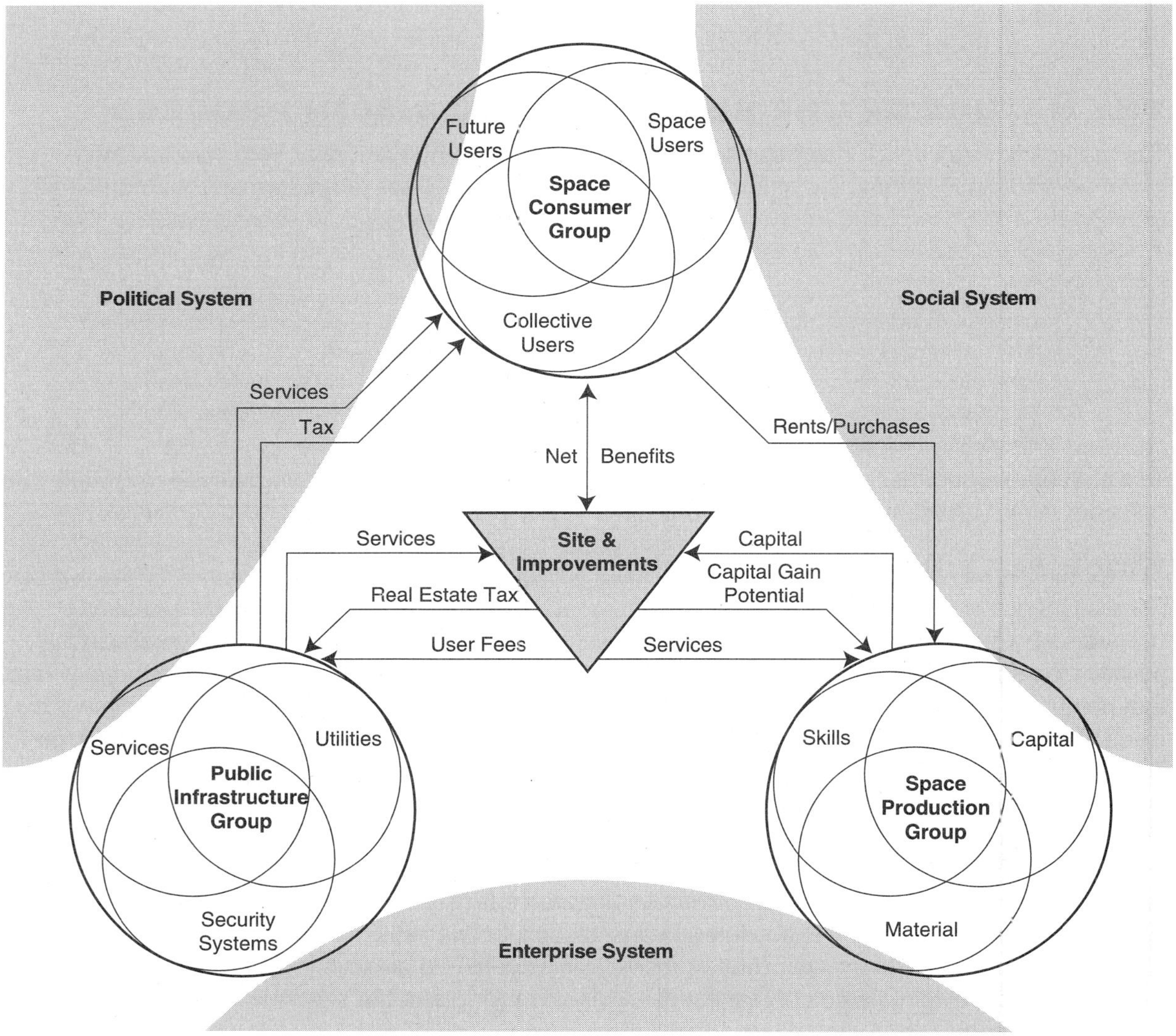

Source: GRAASroot Real Estate Counseling, Inc. From James A. Graaskamp's lecture notes and class handouts.

Location is often identified as the critical factor in a site, but few understand that location value is related to the functional needs of the activity occurring on the site, not the site itself. The family unit dictates multiple locational functions involving employment, school, shopping, and recreation. A family chooses a home site that balances convenience against the cost of inconvenience. Each connection between a household and another point requires movement of persons, goods, or messages. This is called a *linkage* and the time, stress, and dollar costs involved in the movement are referred to as the *costs of friction.*

Each land user seeks a location defined as a set of linkages that will minimize these costs. As the costs of energy, congestion, and time have risen for commuters and the need for suburban school linkages has diminished, some have taken the opportunity to reduce the costs of friction by trading the house in the suburbs for a condominium downtown. Such choices have resulted in increased rent or selling prices for condominium units. Rent differentials for location reflect market recognition of the perceived costs of friction associated with desired amenities. Therefore, locational value is in the mind of the space user rather than inherent in the land, and demand pressures on land shift as perceptions of convenience shift.

Of course, the movement of goods, services, and people often necessitates networks of pipes, paving, and wires connected directly to the site, so some linkages become physically set due to economies of scale in reducing the costs of friction. However, most linkage relationships are subtle, systematic patterns that require a thorough understanding of behavior and are the basis for marketing.

Linkages, Location, and Cash Cycles

Since location has been defined as an attribute of the site user, rather than a physical attribute of the site, it can be best explained through example. The calculations shown in Table 4.1 relate to a retail store. The store's location depends on its linkages to families in the trade area that have both the income and inclination to visit a particular store or supermarket. The presence of other retail establishments nearby may generate traffic volume and attract customers or may intercept customers from the potential trade area. The potential linkage represented by a flow of passing customers can be subtly strengthened by a right hand turn lane or devastated by a median strip that makes a particular store site inaccessable to traffic going by. A great number of linkage relationships will affect potential sales of the store, which, in turn, control the acceptable rent levels, capital budgets, and store sites in the real estate process.

Table 4.1

Calculating Locational Value of a Retail Store

I. Determination of Optimum Store Rent:

	9,400	Number of families in area
×	0.55	the store's capture rate
	5,170	Number of families visiting the store
	$32,500	Average family income in area
×	0.11	% family income spent in supermarkets
	$18,482,750	Total potential sales for the store (family spending x 5,170 families)
×	0.65	Leakage of food purchases to other sources
	$12,013,788	Expected potential sales for the store
÷	$400	Sales per sq. ft. of floor area per year
	30,034	Optimum building size
	$12,013,788	Expected potential sales for the store
×	0.035	% of sales allocated to rent expense
	$420,483	Net rent allowed per year
÷	30,034	Optimum building size
	$14.00	Optimum net rent/sq. ft.

II. Determination of Optimum Building Cost:

	1	Parking stalls required per 300 sq. ft. of *GLA*
	42,925	101 stalls * 425 sq. ft. each
+	34,130	Gross building coverage (30,000/0.87899, say 0.88)
	77,055	Minimum site area
×	$8.00	Price of land per sq. ft.
	$616,440	Price of site
×	0.10	10% interest on land
	$61,644	Land contract payment
	$358,339	Annual budget for improvements
÷	0.135	Capitalization rate for improvements
	$2,658,063	Total budget for building and site improvements
÷	34,130	Gross building area
	$77.88	Building budget per sq. ft.

As the table shows, a store's capture rate has a direct impact on the optimum rent that can be charged. The optimum rent, in turn, has a direct impact on the supportable (justifiable) building budget. Thus, the investment and lending security of the retail property is not the terms of the lease but rather the customer flow into the property. A retail property without customers has no value, regardless of the name on the lease.

If the capture rate for the retail property can be increased, through reduction of the costs of friction, the property's value will also increase. However, the reverse condition is also true—the property's value will decrease if the costs of friction increase. Consequently, a significant part of the property underwriting process focuses on the identification and management of the market elements that affect the costs of friction in the eyes of the property's customers.

Feasibility Analysis

Feasibility analysis is a general term that covers a variety of predevelopment studies by generalists and specialists undertaken to determine facts that are reliable, assumptions about the future that are consistent with past experience, and tactics that will minimize the variance between objectives and realizations. A developer builds only what can be financed, and lenders should finance only those projects for which there is a defined consumer group representing effective demand at a specified period of time.

The sequencing of the analysis depends on the problem. Ultimately, only three types of real estate feasibility problems exist:

1. The search for the most fitting use for a specific site (Figure 4.2)
2. The search for the most fitting site for a use (Figure 4.3)
3. The search for the most suitable investment by investors (Figure 4.4)

The most common problem is the search for the most fitting use for a specific site by a speculative developer (Figure 4.2).

Approaches to Determining Feasibility

The revenues of a development come from either rental income or from the sale of space-time units. The real estate development intended for rental purposes provides the clearest and simplest demonstration of how a space-time unit and a corresponding monetary value are irrevocably linked together. If the total capital budget has been established based on a completed architectural program, it is then possible to determine the rent required per space-time unit using a "front door" approach. More realistically, the developer determines the market rents and additional supply of space required in a given sector and then works backwards using a "back door" approach. This latter approach establishes the capital budget justified by expected revenues and the design is limited accordingly. Too often, the design specifications are set so that the rents required to justify the project are out of reach of the rental prospects in the marketplace.

MARKETING—THE KEY TO DEVELOPMENT

Market Research

All assumptions for raising capital to build real estate rest on revenue flow, which begins with the customer. Selling is a method of persuading the customer to accept what is being sold, while marketing helps shape the product and the service to the needs of the user. In real estate, marketing must serve the needs of three ultimate user groups:

1. Individual purchasers or tenants
2. Collective users operating through the political process to approve or disapprove a project proposal
3. Future users who can be expected to convert a structure to conform to changing lifestyles and needs at some point in the future

Figure 4.2

Feasibility: Site in Search of a Use

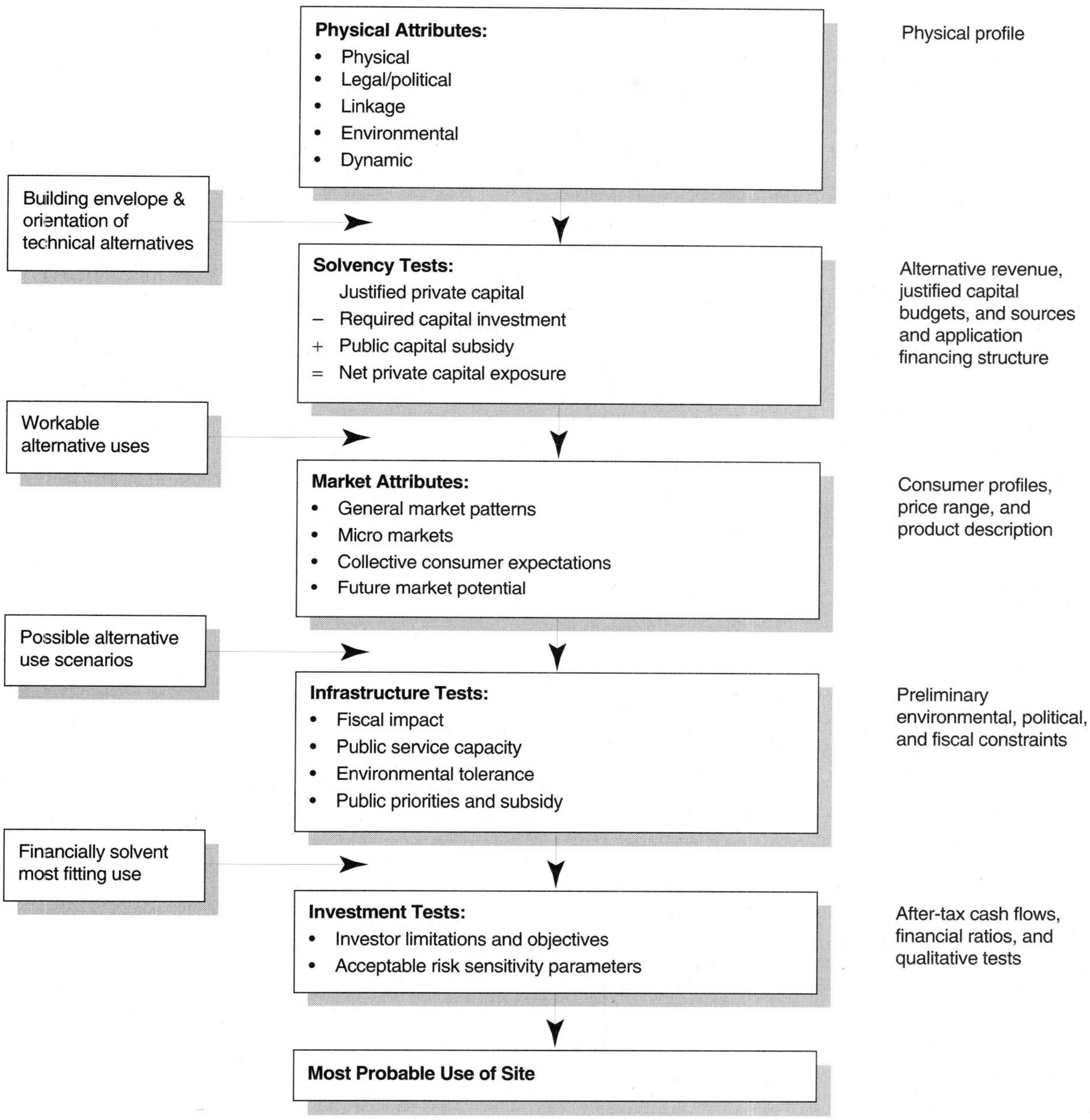

Real estate marketing is directed toward creating a product that is uniquely sensitive to the needs of the consumer and reduces other costs to the consumer to a point where the intended buyer or tenant will pay full price.

A strong marketing position requires careful research into the needs and budget limitations of the prospective user. No American industry spends less on consumer research and product development than the real estate industry. This failure to spend even 1% of a project's budgeted cost on primary research about the intended consumer is one reason why many real estate projects result in business failures. Developers simply misjudge the number of consumers in the market, the needs of those consumers as they perceive them, and the rate at which new units can be absorbed in the marketplace.

Figure 4.3

Feasibility: Use in Search of a Site

Use location profile

Profile Attributes:

- Linkages
  - To markets
  - To employees
  - To supplies
  - To ancillary services
- Legal/political limitations
  - On use
  - On site
  - On investors
- Site functions and size
- Neighborhood and community requirements
- Improvement functions and size
- Environmental impact of activities

Definition of site search parameters ➤

Preliminary rejection of majority of alternatives

Preliminary Screening of Alternatives with Use Profile Criteria:

- Qualitative criteria
- Quantitative criteria (nonfinancial)
- Subjective preferences

Acceptable physical sites ➤

Preliminary financial impact of acceptable alternatives

Financial Impact Forecasts:

Revenue

- Sales revenue
- Operating costs
- Labor costs
- Occupancy cost
- Management cost
- Net income
- Less:
  - Debt service
  - Taxes

Net cash return

Capital

- Net land investment
- Net tenant improvements
- Net building investment
- Relocation cost
- Less:
  - Government grant
  - Subsidized loan
  - New debt
  - New capital

Net capital exposure

Financially viable and workable sites ➤

Decision matrix, scoring systems, or game plans applied to final selection

User Trade-Off Values:

- Capital efficiency vs. employee security
- Marketing edge vs. raw material sources
- Community obligations vs. company efficiency
- Location captial vs. future advertising expense
- Current business practice vs. long-term changes in technique
- Etc.

Ranking of alternatives for acquisition ➤

Selection of Most Probable Site(s)

Figure 4.4
Feasibility: Investor in Search of an Opportunity

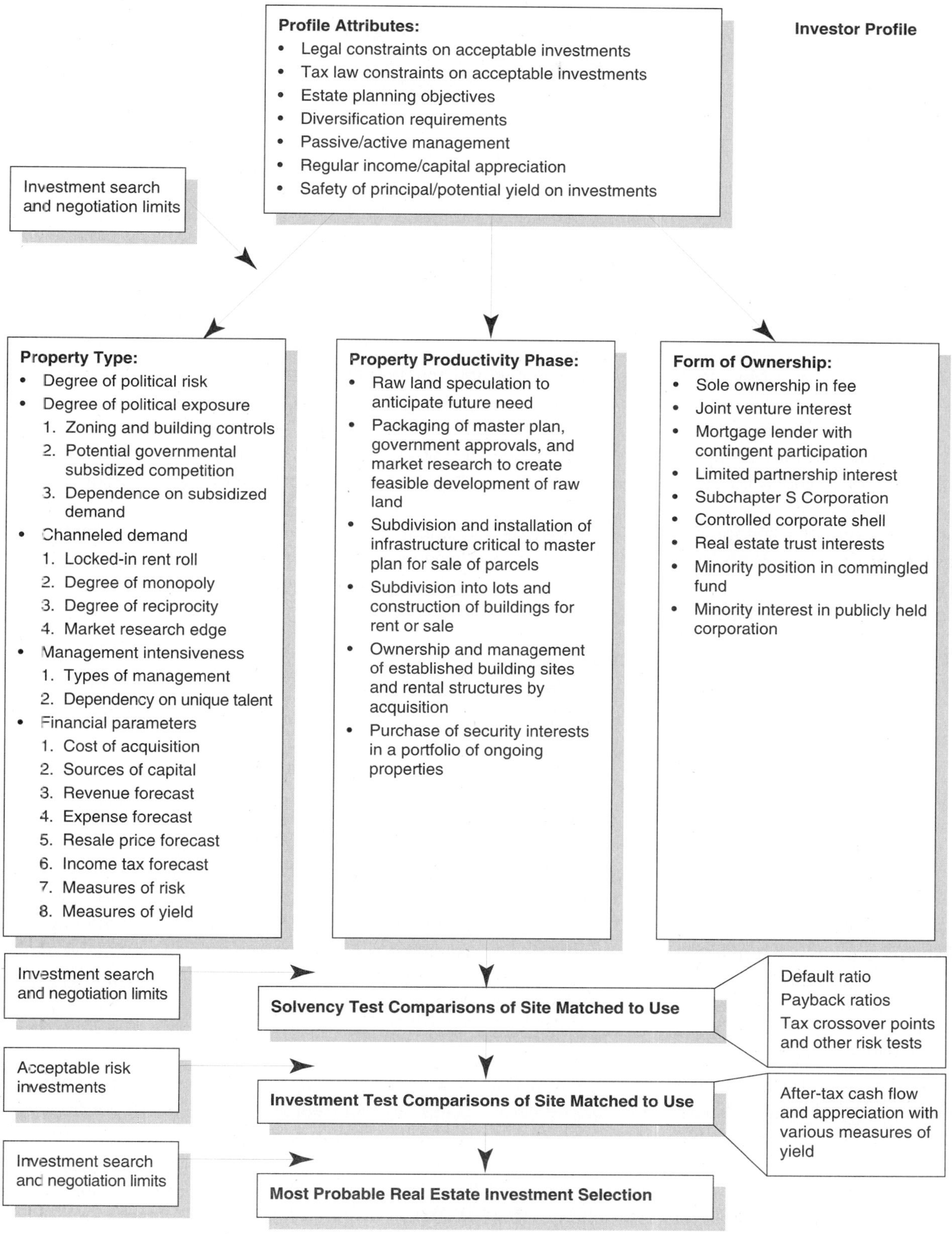

It is a great irony that mortgage lenders expect insurance premiums to be paid in advance in case the project should burn down, but typically require no original research to discover whether the project will rent up. The real threat to the security of their loans is a lack of customers, not the risk of fire. Lenders assume that a developer with a proven track record knows the market from past projects, but past success is no guarantee that any demand remains unsatisfied or that market action has not shifted its location and product preferences. Design serves society best when it serves the intended user, rather than the normative standards of the designer, developer, or yesterday's market.

Market Data Versus Merchandising Data

Market data are typically aggregate data describing a population by age group, income category, business activity, location of residence, average home price, and other statistical information. Such information is collected by the U.S. Census Bureau, regional councils of government, local planning offices, and other public and private sources. Market data will assist in determining the scale of a project so that 800 housing units are not planned for a community that will only need 600 in the foreseeable future. Market data may measure the expected rate of absorption for broad categories of space, say, 500,000 square feet of Class A office space, 200 condominium units, or 600 single-family lots per year. More to the point, however, is how much of that potential market can be captured by a specific project and how the project's capture rate is related to merchandising data.

Merchandising data are typically collected with a specific project in mind. The analyst collects merchandising data through primary research with the objective of determining the *competitive standard* of project attributes and discovering the unmet needs in the market that could provide a *competitive edge.*

A competitive edge depends on finding a truly unmet need on the part of a particular consumer group. It is not obtained through gimmicks or cleverness on the part of the designer or the developer. A competitive edge typically is created by finding ways to enhance user self-esteem or to reduce the costs of friction, anxiety, or inefficient space layouts housing the user's activity.

The Ethics of Fit and Monopoly

There is a direct relationship between these factors:

- The ethics of fitting a project to the environmental constraints of the site
- The needs and budget of the eventual user
- The concerns and fears of the collective consumer
- The uniqueness of a project, which creates a monopolistic dimension in its pricing

A full price that is willingly paid by the individual user indicates satisfaction and maximizes investment value by stabilizing many of the critical risks of the development process for the investor. Feasibility is defined as measuring the fit of the project to its physical context, to its intended users, to the objectives of the investor, and to the limited resources of the developer and the community. While it is taken for granted that the project design will fit the soil profiles and topography of the site (or the structure will collapse or sag), it is not generally understood that the project will financially collapse if it is insensitive to customer profiles and cash cycle topology.

The Role of GIS

The feasibility charts shown in Figures 4.2, 4.3, and 4.4 illustrate the complex and interrelated nature of modern real estate decision making. The nature of the product (real estate's spatial dependency), its vast and complex data structure, its need to integrate past trends and experience (a retrospective viewpoint) with future expectations and assumptions (a prospective viewpoint), and the need for short analytical time frames all argue for the use of GIS. In modern real estate decision making, the "what" of the problem statement must be combined with the "where" of the data structure—and a GIS is uniquely able to assist in performing this task.

GENERAL TYPES OF SPATIAL DATA

Data within a GIS are organized as separate layers and are stored in two types of databases: a graphics database containing locational information and an attribute database containing descriptive information.

Data in the graphics database are represented by points, lines, and polygons (areas). Within this database, locational data may exist in two structures: vector or raster. Vector data structures preserve the format of traditional map features represented as points, lines, and polygons with (x,y) coordinates; the usual coordinate system is longitude-latitude, but numerous others such as the state plane coordinate system also exist. Data in a raster format are placed into an array of grid cells identified by a row and column number. A variant of the raster structure is quadtree. In a quadtree structure, a region is divided into smaller and smaller quadrants, depending on the level of detail required to describe the map feature. Once a quadrant is completely homogeneous, subdivision ceases.

The attribute database allows for storage, manipulation, and reporting of attributes. Attribute information may exist in three types of databases: hierarchical, network, and relational. The relational type of database is the most popular.

Spatial Measurement

The most basic function of a GIS is the calculation of distances between geographic entities and the measurement of areas. A GIS can calculate the distance between an office building and an interstate highway; it can also compute the area of a polygon such as the trade area of a shopping center. These spatial measurements can then be used in familiar econometric models, which typically include approximate locational measures.

Figure 4.5 is a map of a sampling of the major apartment projects in Denver County, Colorado, combined with noise contours from Stapleton International Airport. With these two data items, one can measure the distance from any located apartment project to a stated noise level (e.g., 75, 70, 65, or 60 dBs).

Generation of Buffers

The generation of buffers with a GIS allows the user to either exclude or include entities that are within or beyond a specified distance. Buffer maps may consist of a set of concentric circles, in the case of point entities, or corridors, in the case of line entities. For example, the buffering function of a GIS can be used to identify the apartment buildings that are less than two miles away from an interstate highway and less than four miles away from a shopping mall.

Graphic Area Data

The basis for most area-based socioeconomic analysis is the U.S. Census TIGER\line files. TIGER (topologically integrated geographic encoding and referencing system) is the census bureau's digital map database, which automates mapping and related geographic activities required to support the census and survey programs. TIGER provides coordinate-based digital cartographic information for the entire United States, Puerto Rico, the Virgin Islands, and the Pacific territories over which the United States has sovereignty or jurisdiction.

The TIGER database includes

- Physical features such as roads, railroads, and rivers; feature names and classification codes; alternative feature names and codes; and feature shape coordinates
- Address ranges and zip code zones within metropolitan areas
- Census statistical area boundaries such as census blocks, block groups, and census tracts
- Political boundaries of states, counties, and incorporated places

Since the 1990 census, the census bureau has periodically released updated TIGER files, and various commercial venders have enhanced the records further. The TIGER data from 1990 and the

Figure 4.5
Airport Noise Contours and Apartment Locations

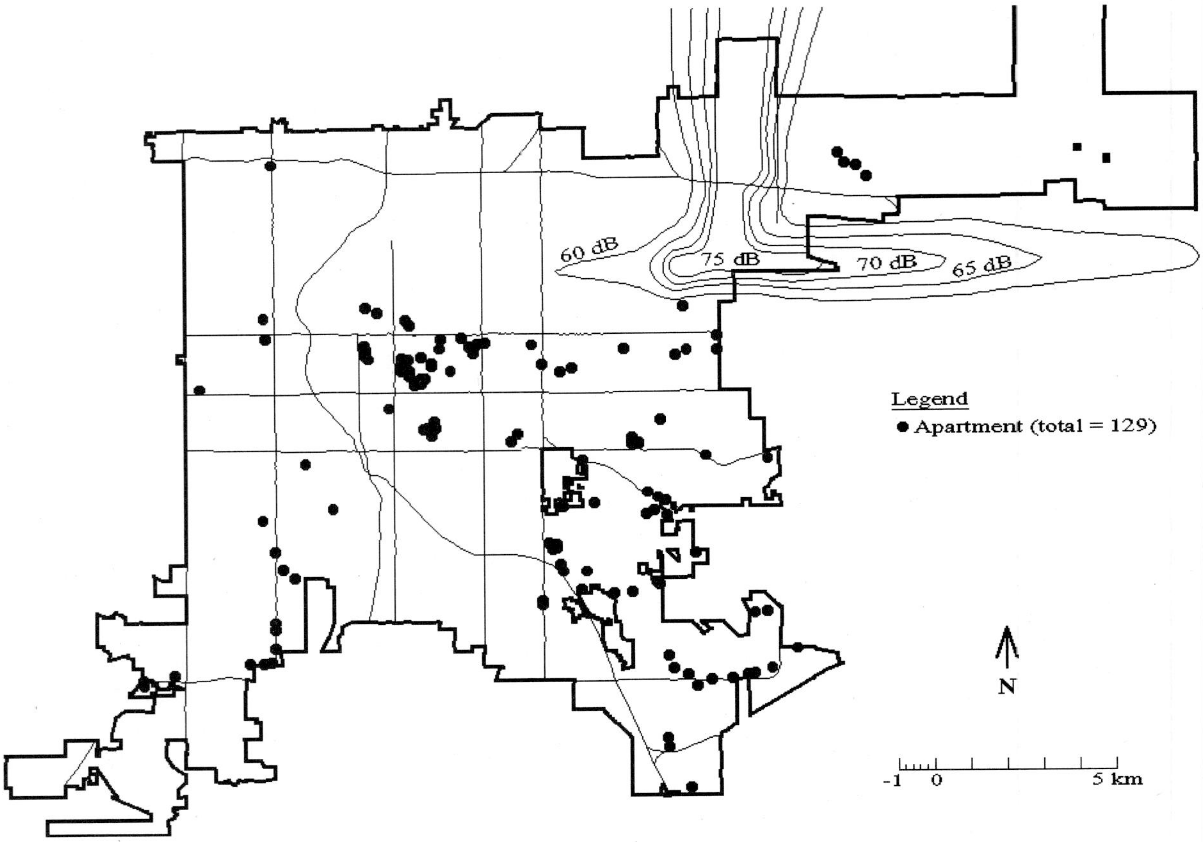

updates are available at most public libraries. The CD-ROM disks contain not only the data but also software for display of the data.

The census bureau has established three commonly used units of area coverage:

Census block Census blocks are small areas bounded on all sides by both visible features, such as streets, roads, streams, and railroad tracks, and invisible boundaries, such as city, town, township, and county limits; property lines; and short, imaginary extensions of streets and roads. In order to ensure confidentiality, the public is not generally permitted access to tabulation data at the block level of geography.

Block group A geographic block group is a cluster of blocks within a census tract that share the same first digit in their three-digit identifying numbers; that is, several block groups comprise a census tract. Block groups generally contain between 250 and 550 housing units, with an ideal size of 400 housing units. The block group is generally the smallest unit of geography for which census data are available to the public.

Census tract Census tracts are relatively permanent statistical subdivisions of a county. Census tracts are delineated for all metropolitan areas and other densely populated counties by local census statistical area committees following census bureau guidelines. Census tracts usually have populations of 2,500 and 8,000 and, when they are first delineated, are designed to be homogeneous with respect to population characteristics, economic status, and living conditions. The spatial size of census tracts varies widely depending on the density of settlement. Census tract boundaries are delineated and maintained over a long period of time so that statistical comparisons can be made from census to census. However, census tracts occasionally are split due to large population growth or combined as a result of substantial population decline.

Attribute Data to Match with Graphic Area Data

The point, line, and polygon data files generally do not provide enough information for analysis. Having mapped the block groups or census tracts contained within a study area, an analyst begins to evaluate various conditions pertaining to population, housing, or income characteristics. To facilitate this task, the census bureau provides summary statistics that describe characteristics at the levels of geography defined in the TIGER structure. These data can be purchased on CD-ROM from the census bureau for $250 each (population, housing, and income data) or can generally be accessed at the local public library.

In addition, one can purchase the census bureau's data and enhancements from various commercial vendors. These vendors specialize in selling enhanced and updated levels of public record information. Vendors such as CACI, Claritas, and Market Statistics specialize in updating and distributing various combinations of socioeconomic data. (Chapter 7 of this text provides more information on data sources.)

Point Data

Point data do not have any area, only a location. An office building, for example, can be represented by a point identified by a single *x,y* coordinate pair and a descriptive label such as "office building."

In real estate a broad spectrum of data begins as point data. For example, the street address of a property is point data. The address generally identifies a specific parcel but provides no indication of the property's size or condition. To find a property the user must first identify the subject's street address and locate it on a map. Then a plan of action is formulated based on where the property is located and what is located around it. This process of locating a street address on a map has been automated and is generally called *address matching* or *geocoding*.

The process of expressing a street address as an *x,y* coordinate pair is a very powerful tool in real estate analysis. Geocoding is an efficient way to convert a well-established common referencing system, the street address, onto a spatial system suitable for automatic mapping and display.

Attribute Data to Match With Point Data

Having located a property on the map, the next step is to examine the features associated with the parcel. In many instances this begins with the public record, most notably the tax assessor's inventory of the property. The tax assessor inventory describes the "attributes" associated with the point. Lot information such as size, shape, topography, and view can be examined as can improvement data such as size, age, condition, and style. Some real estate data vendors (e.g., Experian, DataQuick, COMPS Inc.) repackage such information for more convenient use.

Point to Area Data

In many real estate analyses, it is necessary to define areas of similar attribute coverage. For example, it may be informative to identify the "neighborhood" of the subject property in terms of a similar level of a specified characteristic, such as "market value." A color-coded dot map could be prepared in which the colors of the dots indicate the market value category of each mapped property. Interpretation of the pattern of the dots could be valuable but might be visually difficult. Visual interpretation of the resulting pattern could be greatly enhanced, however, by linking together all properties falling into the same market value category and painting all intervening space with a single color. With this method, point data are converted into area data and the "neighborhood" of the subject property can be better defined. Moreover, area calculations can be performed to determine various units of measure for more precise identification and comparison.

CASE STUDIES

Improved Property Analysis: Bear Valley Area

The improved property analysis project described below was sponsored by Marshall & Swift Valuation Services. The purpose of the project was to examine the usefulness of combining replacement cost procedures with automated sales comparison techniques in an attempt to identify the spatial distribution of casualty risk for the insurance industry.

The Bear Valley neighborhood is situated in the southwest corner of Denver County, Colorado, approximately six miles from Denver's central business district. Figure 4.6, which shows the county boundary, census tracts, and streets, was prepared by layering each of these features in a GIS (SPANS MAP) using data from the 1990 TIGER file for Colorado.

The Bear Valley neighborhood was developed as a single-family, detached housing estate in the mid-1960s and is typical of developments of that era. The neighborhood is 3.5 square kilometers in area and contains two census tracts, five census block groups, 71 census blocks, 1,540 housing units, three schools, and two parks.

Figure 4.7 illustrates distance buffering. In this instance the distance to the nearest fire hydrant is shown for a specific census block group.

Figure 4.8 is a point map of the residential housing units. It was created from a list of property street addresses using address matching, or geocoding, in which a longitude-latitude coordinate was automatically determined for each property. In addition to locating the properties' approximate position, the dots indicating each property are color-coded for the data attribute of square footage of living area.

Figure 4.6
Layered GIS Map

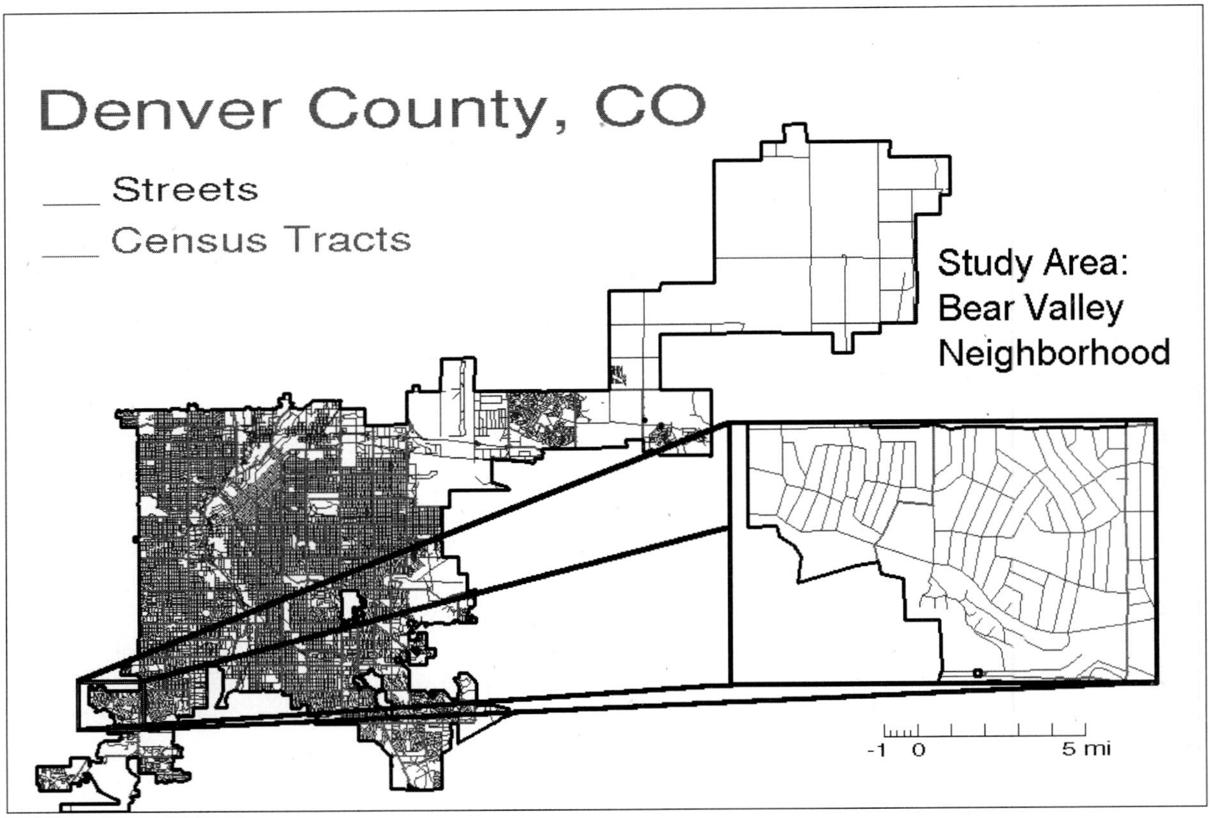

Figure 4.7
Bear Valley—Hydrant Location

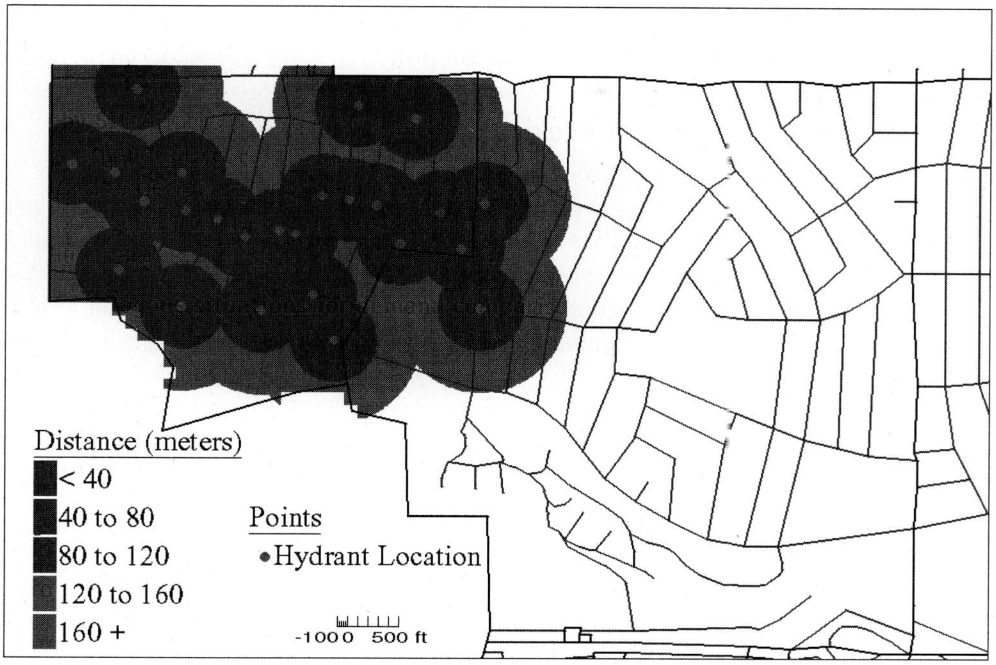

A color copy of this figure can be found on page 185.

Figure 4.8
Bear Valley Neighborhood

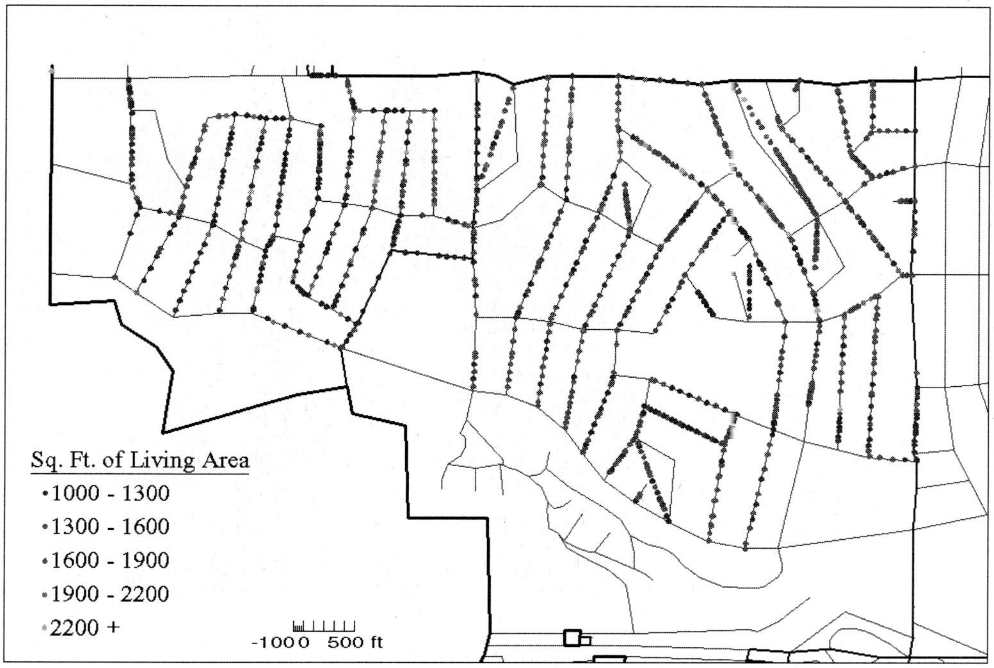

Figure 4.9 demonstrates the presentation power of the mapping programs, with maps, spreadsheets, and charts displayed simultaneously on a computer monitor. The figure includes

- A point map with the points color-coded to identify replacement cost categories
- A spreadsheet table indicating the address of the highlighted point
- A chart illustrating the difference between the property's current assessed value and its replacement cost
- A scanned, bit-mapped image of the neighborhood atlas
- A window for viewing the replacement cost report

In this example, the replacement cost report is physically located not in the GIS property attribute database but in an entirely separate database. The GIS property attribute database contains only an index reference to the replacement cost report database.

Figure 4.10 illustrates the combination of maps, video images, and reports from two independent estimates of value—the cost and sales comparison approaches. In the figure, four active windows are open and visible:

1. The map window in the upper left corner displays the TIGER streets and residential unit locations. In addition, residences within a specific census block that have sold in the past two years are identified with a red dot and outer ring.

Figure 4.9

Presentation of Maps, Chart, and Spreadsheet

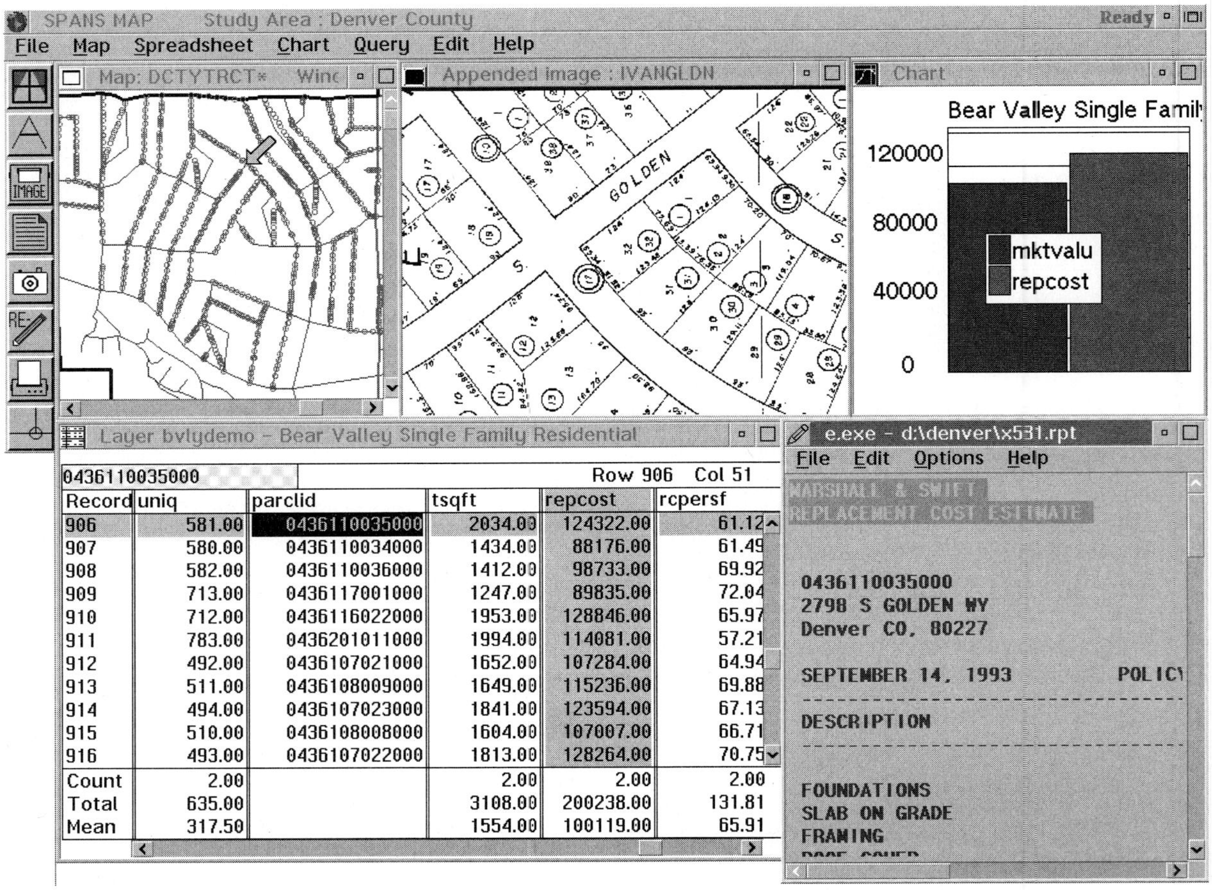

A color copy of this figure can be found on page 186.

Figure 4.10
Presentation of Cost and Sales Comparison Data

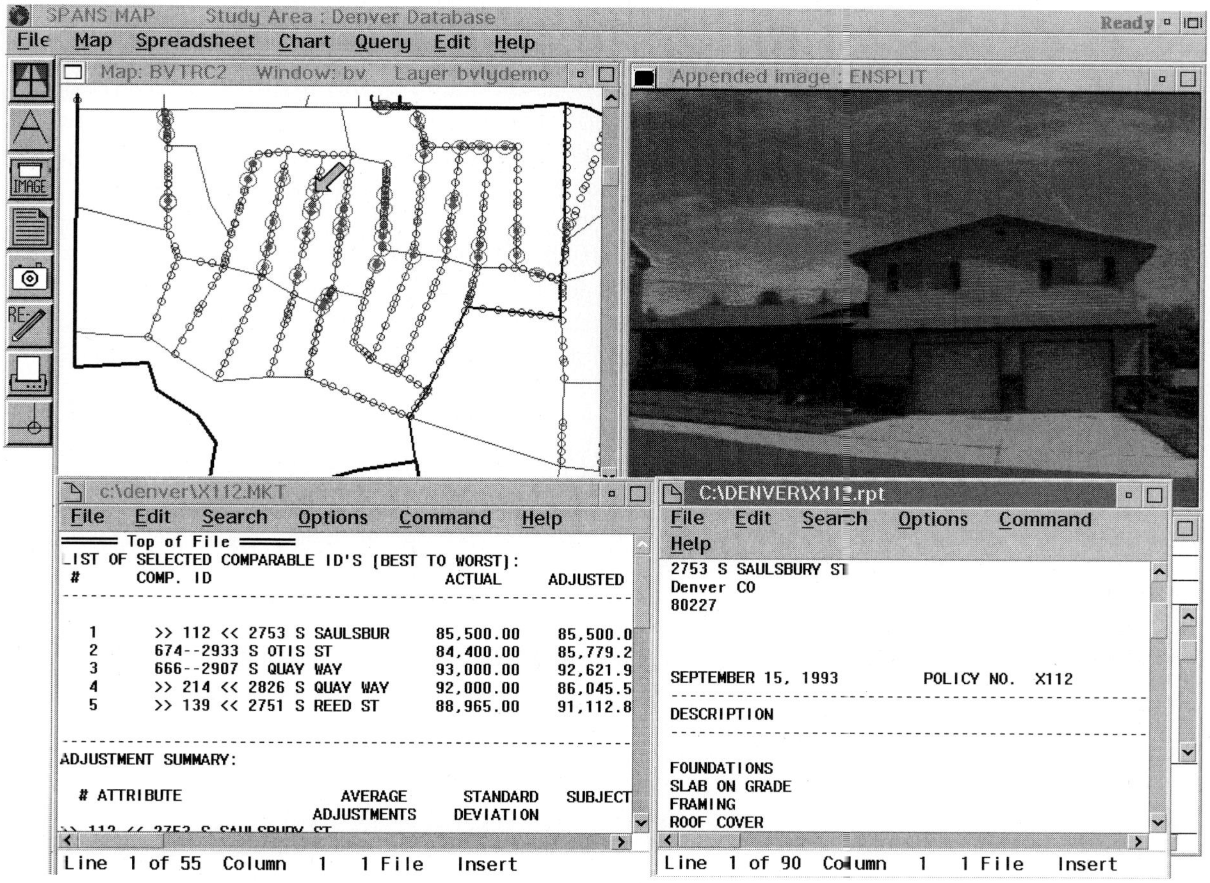

A color copy of this figure can be found on page 187.

2. The window in the upper right corner displays a video image of the property being pointed out on the map window.
3. The window in the lower right corner displays a portion of the first page of the cost report.
4. Finally, the window in the lower left corner displays a portion of the first page of the sales comparison approach report.

For the most part, the data used to create these displays are readily available throughout the nation. If the property attributes are not available directly from the tax assessor, they can be provided by public vendors such as DataQuick and Experian. The street files and socioeconomic data can be borrowed from the local library, or enhanced data can be purchased from various vendors. The video images were created with a standard video camera and loaded onto the computer with an inexpensive video capture software product. The value estimates were created with in-house software (the sales comparison approach) and with commercial software from Marshall & Swift (the cost approach).

Unimproved Property Analysis: The Morrison Quad

The valuation of unimproved land is a rapidly evolving area of real estate analysis. The ability to apply very sophisticated analytical concepts through the use of inexpensive data structures and software holds enormous promise for GIS enthusiasts. Before a determination as to the suitability or capability of unimproved land can be made, the modern real estate analyst must understand the interplay of the property's

- Surface characteristics (percent of slope, slope aspect, and relative elevation)

- Subsurface characteristics (depth-to-water, depth-to-rock, pH, etc.)
- Characteristics above the surface (view from the property, view of the property, sun exposure, drainage patterns, noise, air quality, etc.)

To support this complex decision-making process, an inexpensive, readily available data structure known as digital elevation model (DEM) data is available from the U.S. Geological Survey (USGS). DEM data consist of a matrix of elevation points produced by the USGS as a work product in preparation for the production of topographic maps. The examples presented in this case study make use of the DEM data for the Morrison, Colorado, quad sheet at a scale of 1:24,000. The map and DEM data cover an area of approximately 55 square miles, or approximately 6.5 miles by 8.5 miles. The area contains approximately 35,200 acres.

The software used to produce the following exhibits is MICRODEM, which was obtained at a nominal cost from Dr. Peter Guth. Many other GIS packages can perform similar types of analyses.

Figure 4.11 is a sampling of the data matrix created by the MICRODEM program to support the terrain mapping and analysis. The original DEM data, as purchased from the USGS, had a 30-meter density; with one data point every 30 meters, the total number of data points for the Morrison quad is approximately 675,000. Using this *x,y,z* matrix of elevation points, several straightforward analyses can be performed. For example, Figure 4.11 is also a simple contour map where points of equal elevation are connected with a line.

In addition to studying the distribution of relative elevation ranges, the real estate analyst is concerned with other conditions of the terrain. Figure 4.12 is a color display of percent of slope: the green-to-gray colors represent land that is relatively flat (less than 15% slope), while the brown-to-red colors represent land that is relatively steep (greater than 25% slope). Percent of slope is very important in assessing the suitability of a parcel for a particular use. A parcel of land can be too flat

Figure 4.11
Sampling of Data Matrix

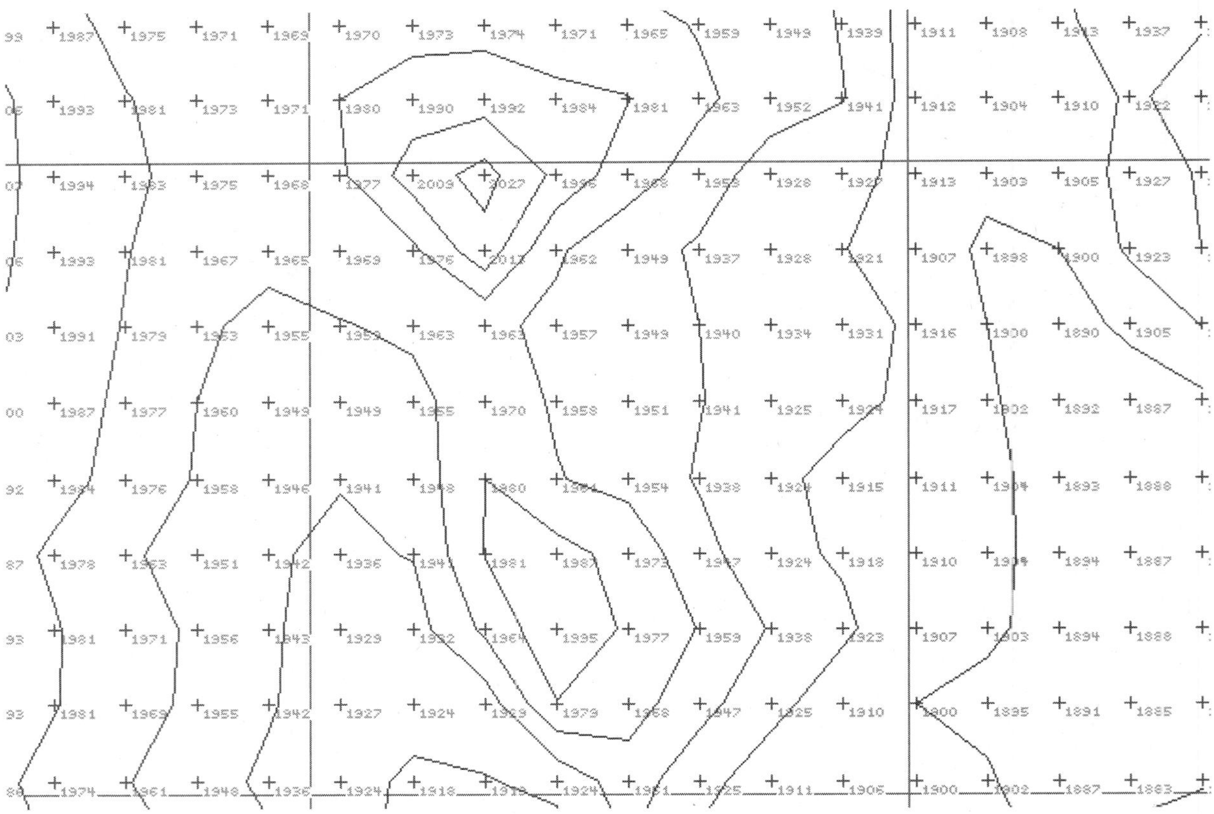

Figure 4.12
Percent of Slope

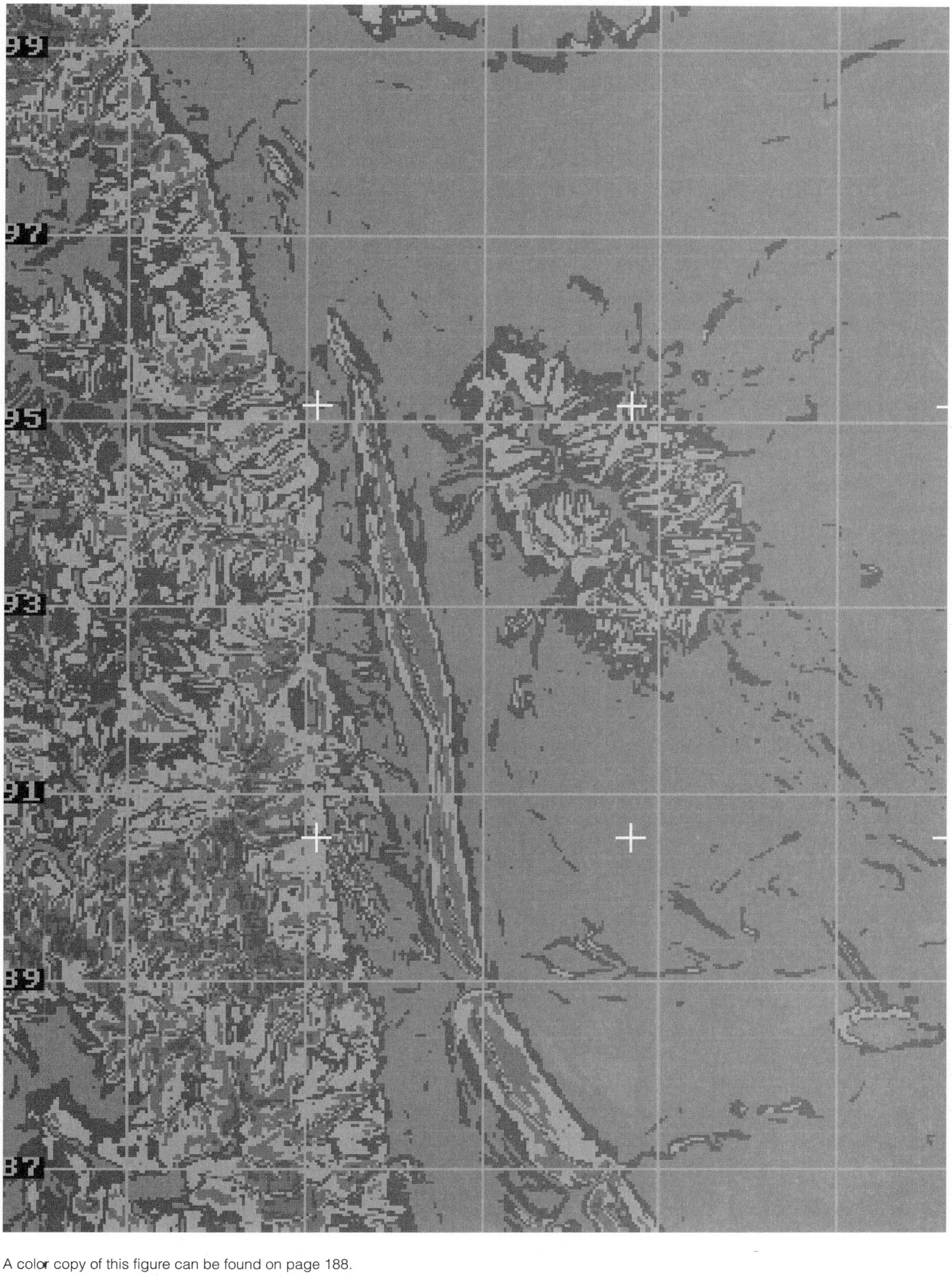

A color copy of this figure can be found on page 188.

or too steep and either condition will require additional expenditures for drainage control and modification. The extent of modification required, especially when combined with surface and subsurface soil and rock conditions, can quickly render a site unsuitable for a particular use.

Figure 4.13 displays another important site characteristic—slope aspect. Slope aspect is the direction that a slope is facing. The green color represents land with a southerly facing direction, while the blue color indicates land with a northerly facing direction; the other colors represent land with either an easterly or westerly facing direction. In many areas the slope aspect of a site will determine the site's view, sun exposure, and micro-climate conditions. These last two characteristics directly affect heating and cooling costs as well as water demand and vegetation suitability.

Figure 4.14 is an illustration of a photo-realistic visualization technique commonly referred to as *shaded relief mapping.* Automated shaded relief mapping is based on a model of what the terrain might look like if it were made of an ideal material and illuminated from a given position. The final results resemble an aerial photograph because of the use of gray scales and continuous tone techniques of display.

Figure 4.15 shows an enlarged portion of the Morrison quad. This area has been detailed with contour lines of equal elevation, overlaid with both UTM coordinate lines for spatial reference and with the local TIGER road network. Finally, the figure has been used to conduct a *viewshed analysis*.

To prepare the viewshed analysis, the location and elevation of the observer were identified. In this case the observer is two meters above the ground surface, approximating the height of a 6-ft. person. From the specified position, a *view fan* was identified. In this example the view fan extends from an approximately southeasterly direction around to approximately due west. Then, beginning at the southeasterly edge, a series of intervisibility lines were determined from the point of origin (the observer) and extended outward to a maximum distance of 3,000 meters. Along the intervisibility line a series of calculations were performed to determine if the point of the landscape would be visible to the observer. When the point was visible to the observer, the point on the landscape was highlighted with a green line segment. When each intervisibility line was finished, the green area along the line represented the land area exposed to the viewer; when all intervisibility lines contained within the view fan were combined, a map of the area visible to the observer was created. This is the viewshed from the observation point.

Case Study Conclusions

The GIS applications described above are only a sampling of the types of analysis that can be accomplished today with a minimum investment in software and data. In most instances the hardware found in a typical real estate office is fully capable of supporting entry-level GIS applications. The true costs of GIS are tied to the training and experience of the user.

As an industry, real estate has not expected much from its representatives. The skill level needed to become a "qualified real estate professional" is low, and in most instances there are few incentives to enhance one's skills. For example, many lenders do not request or expect "market value estimates" from appraisers; they are only interested in "transaction justification estimates." An appraiser who submits value estimates that kill a deal may be shortly out of work. Often developers could not care less if their proposed projects ever rent up; their only consideration is selling the deal, and selling the deal with hype is much easier than doing systematic, insightful research into markets and market demand.

However, the winds of change are blowing. Government regulators and individual investors will remember the crash of the mid-1980s for a long time. They are beginning to want and expect more from real estate analyses. In all aspects of real estate, from automated valuation systems to investor criteria, modern analysts must become more accountable. GIS provides tools with the potential to meet the greater accountability needs of real estate decision making.

Figure 4.13
Slope Aspect

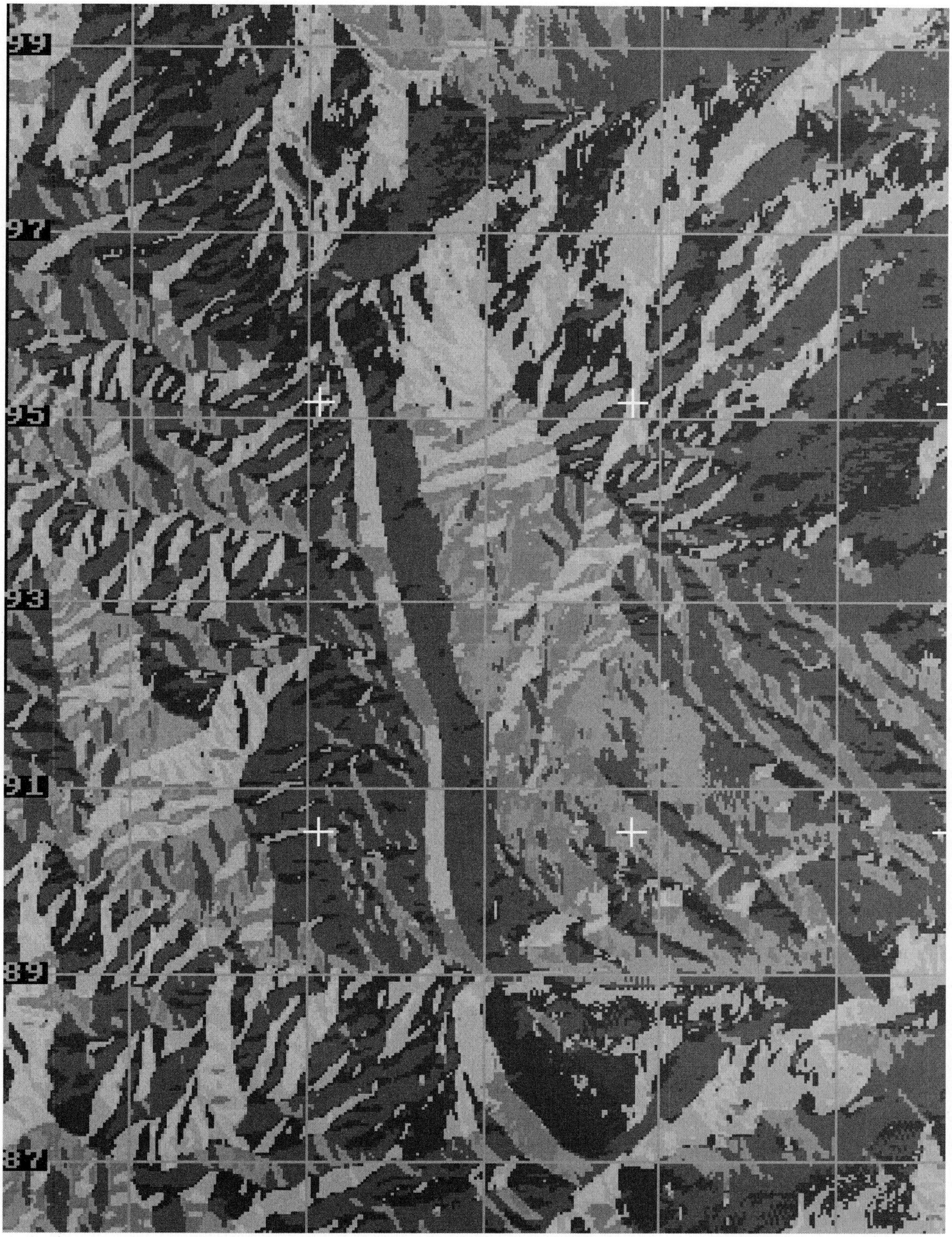

A color copy of this figure can be found on page 189.

Figure 4.14
Shaded Relief Map

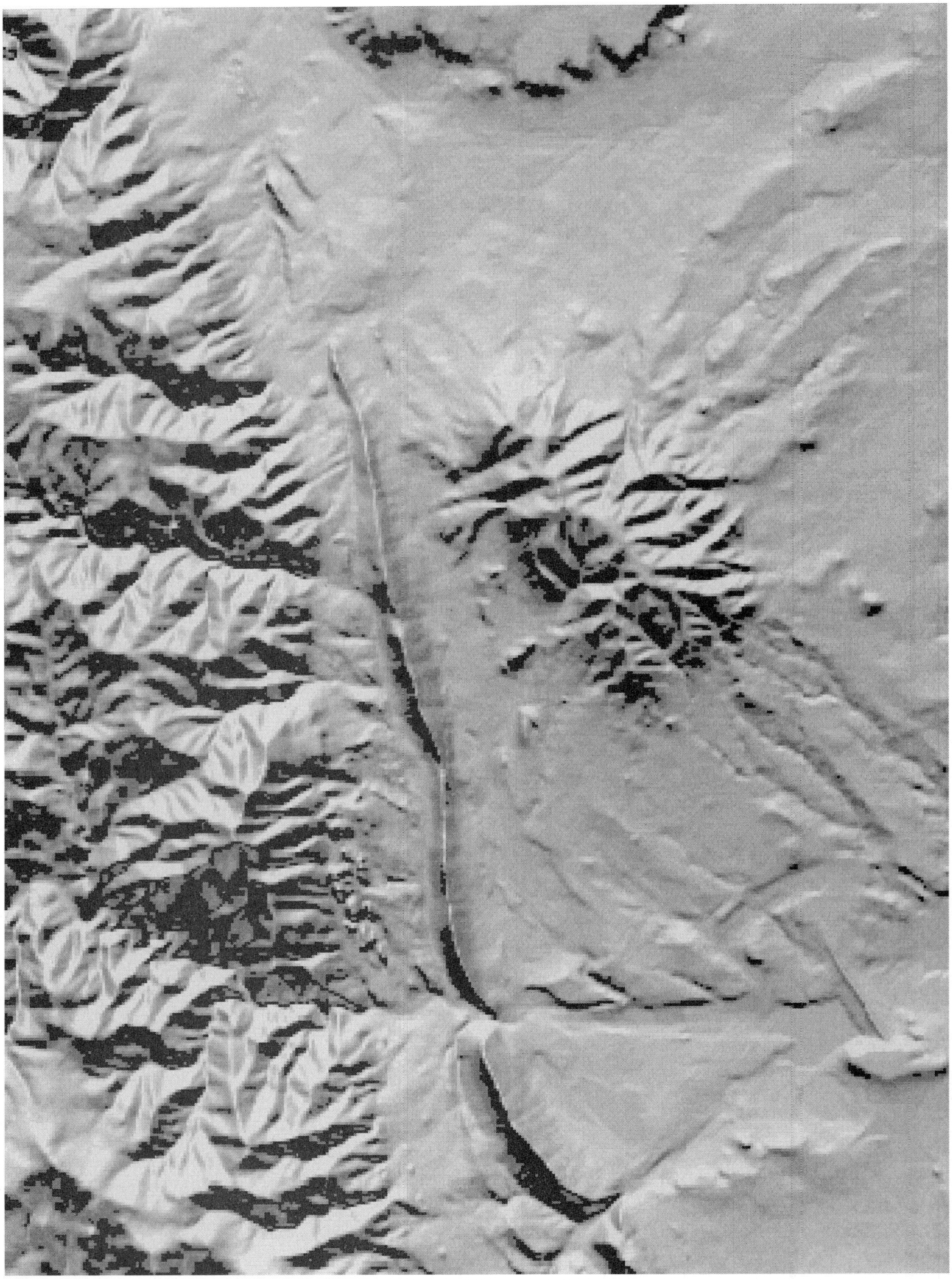

Figure 4.15
Enlarged Portion of Morrison Quad

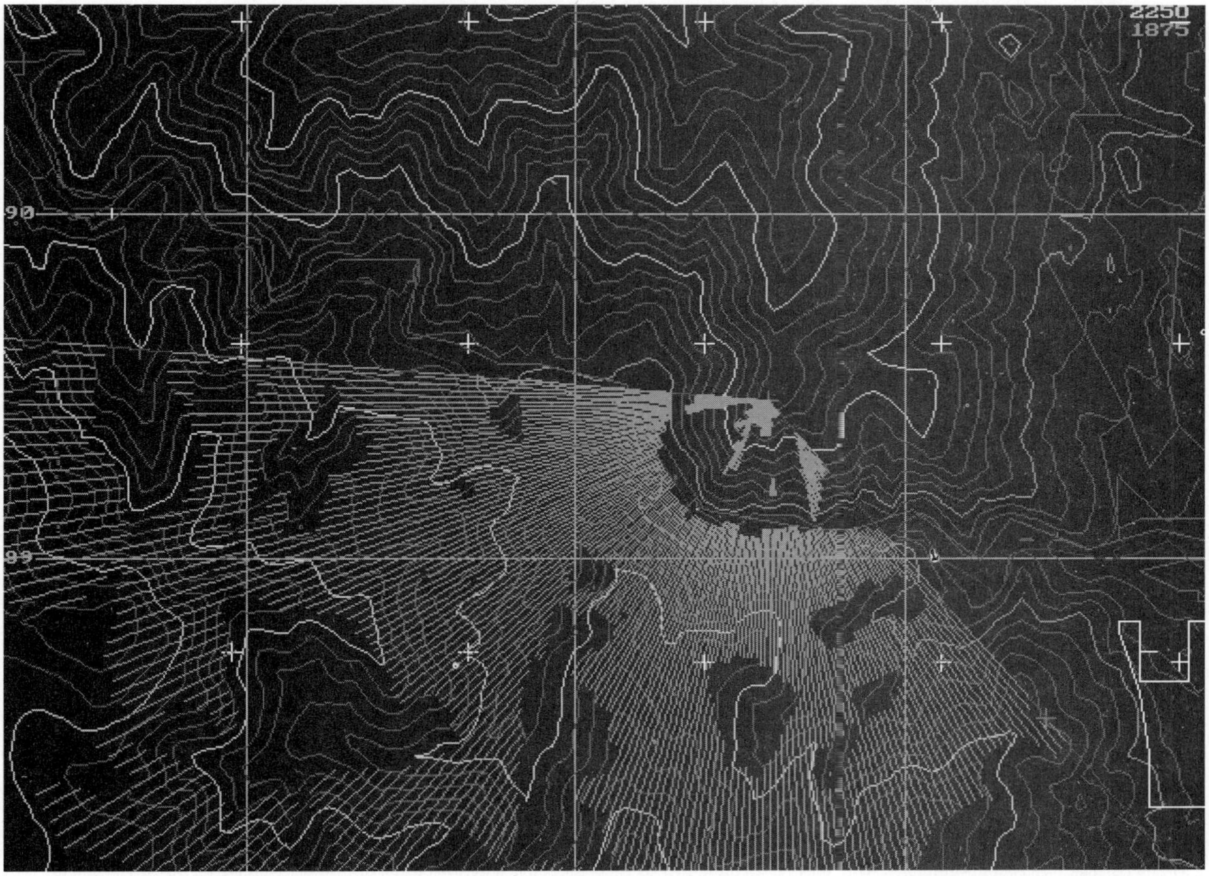

A color copy of this figure can be found on page 190.

SPATIAL DECISION SUPPORT FOR AUTOMATED RESIDENTIAL PROPERTY VALUATION

Purpose

Automated valuation systems are used to reduce the time and cost associated with loan underwriting and to help the appraiser prepare more accurate and supportable appraisals. The objective should not be to replace the appraiser but rather to increase the appraiser's efficiency in rendering an opinion of value. Experience is showing that, almost 80% of the time, the appraiser can render an acceptable opinion of value simply by reviewing the analysis prepared by an automated valuation system.

Valuation Methodology

The reliability of any automated valuation methodology (or any appraisal) that purports to predict the probable selling price of a subject property is dependent on how closely the methodology replicates the decision-making behavior of buyers and sellers. This is the way the market establishes price.

In general, appraisal theory approaches the concept of value by synthesizing three different models of economic behavior. In appraisal, these three models are called *approaches to value* and are referred to as the cost approach, the sales comparison approach, and the income approach. Of the three approaches to value, the sales comparison approach is the one usually applied to value owner-occupied residential real estate. The sales comparison approach to value attempts to use direct market evidence to establish

- What was the value of previous sales, adjusted for economic shifts
- What is the current economic setting (on which to base an estimate of market value)
- What may be the estimated selling price

The analysis of economic behavior is structured from the viewpoint of a typical market participant and is focused on several general categories of data. Improvements to the site are generally subdivided into:

- Numerical data attributes
- Qualitative data attributes

Conditions of the site are generally subdivided into:

- Locational - linkage
- Conditional - environmental (terrain)
- Perceptual - dynamic

Neighborhood conditions around the site are generally subdivided into:

- Social/economic characteristics
- Perceptual - dynamic

Strategic Objectives of Automated Valuation Methodology

In estimating what a subject property will bring in the market, neither automated valuation methodology nor appraisal is concerned with some abstraction called *value* but with a transaction price—the number of dollars that a buyer would pay a seller to acquire title to the property. In appraisal, this anticipated transaction price is referred to as *market value.*

Since the intent is to estimate what price the property would bring if exposed to the market, the methodology must predict the outcome of an event that has not yet occurred, i.e., the hypothetical sale of the property. The analysis is usually a short-term forecast that assumes immediate exposure in the market and as prompt a sale as current market conditions permit. Market value is market-determined—the outcome of the competitive interactions of buyers and sellers, not the value any given buyer or seller might attach to the property in a noncompetitive environment. The value to the owner or to a prospective investor is a subjective value, individually determined and differing among individuals who have different value systems and financial situations. The actual selling price will usually represent a compromise between what the buyer would have paid if necessary and what the seller would have taken as a last resort.

Like appraisal, automated valuation methodology is economic analysis under conditions of uncertainty. Accordingly, its findings can be expressed only in terms of probability. No methodology is capable of making a precise prediction of a probable selling price. At best, the analyst can indicate a range of prices within which the selling price would probably fall; in some cases, there may be sufficient information to venture a judgment as to the probability that the actual price will fall at various points within the range.

In analyzing the value of a parcel of real estate, the starting point is the parcel's inherent utility—i.e., the characteristics and qualities that make the parcel productive and desirable and for which people are willing to pay. However, price is set in the marketplace. To serve the client's needs, an automated valuation methodology or appraisal seeks to predict the price at which the subject property will probably sell. Viewing the property as a package of potentially productive qualities, the methodology must predict the outcome of the interaction of the market forces of demand and supply to which the property might be exposed and which could trigger a transaction. The methodology applied must predict how buyers and sellers will behave with respect to the subject property when it is exposed for sale.

The Working Environment of Automated Valuation

In a typical lending environment, the borrower applies for a loan and the lender reviews the loan application for three critical components:

1. The borrower's job and income status
2. The borrower's credit history
3. The collateral value of the property

With electronic databasing, the borrower's job, income status, and credit history are available to the loan underwriter in less than an hour, usually in only minutes. Thus, two of the three critical underwriting components can be made known to the underwriter during the applicant's interview. Determining the third component, the collateral value of the property, typically requires much more time and cost. The underwriter orders an appraisal, which is generally paid for in full or in part by the borrower. While the cost of the appraisal is not usually a major item of concern, the wait is a source of considerable frustration for both the underwriter and the borrower. Once the appraisal is completed and submitted to the lender, the wait and cost is still not over. The appraisal must then be reviewed by the lender's internal review staff. This step can add another day or more, depending on the review staff's work flow and if the appraisal needs to be sent back to the appraiser for additions or corrections.

With an automated valuation system, the lender's review staff processes the valuation request and, minutes later, the reviewer can be presented with a detailed valuation report, complete with quality scores indicating the overall suitability of the value estimate. With this in hand, the reviewer can determine if any further investigation is necessary. If additional work is needed, the quality scores will indicate whether review, inspection, or reappraisal is required. Next, the data and analysis performed by the software is made available to the property inspector or appraiser, who has been given a head start on the appraisal assignment.

An automated valuation system provides the user with several significant advantages. By using internal staff to first process the valuation request and review the results, the reviewer's knowledge and experience can be put to better use. Significantly less time is needed to process and review the initial valuation request than to review a traditional form appraisal. If the reviewer is satisfied with the value estimate, no additional time or cost is incurred by either the borrower or the lender. When additional work is needed, the field person (property inspector or appraiser) is provided with much better data than they otherwise would have access to, thus the quality of their work is improved and the subsequent internal review is processed more efficiently.

Market Demand

Automated valuation is here to stay. Some lenders are beginning actively to seek out providers of automated valuation systems. Lenders who lack the foresight or resources to create their own systems will be forced to establish relationships with outside venders or will leave the real estate lending business. (Of course, given that real estate is the largest nongovernment business in the United States and is characterized by very large debt-to-equity ratios, any decision to get out of the real estate lending business cannot be made lightly.) While some lenders will elect to give up on the real estate lending business, those who remain will need to compete. A traditional appraisal that costs $400 and takes seven to 10 working days to complete cannot compete with an alternative appraisal that costs perhaps $125 and takes five minutes.

In addition, evidence suggests that when automated valuation systems are used by capable appraisers, the results are significantly more accurate than the results of traditional appraisals. Thus, the combination of reduced cost, faster turnaround time, and increased quality (cheaper, faster, better) suggests that lenders will be drawn to automated valuation systems in the interest of economics and survival.

The Production Product

A GIS-based automated valuation system called AVS is currently being used to value single-family housing. At the core of AVS is a proprietary valuation engine that I have been developing since the mid-1970s. AVS was installed in Seafirst Bank in Seattle, Washington, in January 1995. After eight months of testing by bank auditors and with regulatory review, AVS has been accepted and is now supporting the bank's consumer loan underwriting process.

The proprietary valuation engine is an automated real estate valuation system that combines the efficiency of electronic data processing with accepted appraisal logic and processes to provide supportable estimates of market value for various classes of real estate. The major developmental goal has been to apply the valuation engine to the valuation of single-family, detached residential properties. However, because the program is strongly linked to accepted appraisal theory and processes, AVS can be applied wherever and whenever the sales comparison approach to value is needed, given an adequate number of comparable transactions.

The AVS software product consists of a shell built around a valuation engine, which allows users to perform specific tasks and obtain desired outputs. Experience to date clearly indicates that this shell can be customized to meet special user needs and to adapt to changing data conditions. Nevertheless, the valuation core remains relatively stable and will require alteration only when enhancements are added.

Life Cycles

While the core logic that AVS is built upon is more than two decades old, the combination of data availability, hardware capability, and user acceptance is a very recent phenomenon. Government regulations that raised the de minimis threshold (i.e., the level of debt at which an appraisal is required) to $250,000 in 1994 opened the market to alternative review procedures. This, in turn, spawned serious interest in going beyond de minimis and fully utilizing technology to help lower the cost of lending, in terms of both time and money.

As mentioned previously, AVS was born in the early 1970s with the initial development of the valuation engine. Later, as the use of personal computers in the workplace became widespread, property data became available in an electronic format, and GIS tools became ever more easily integrated with other types of software, the stage was set for the creation of a GIS-based automated valuation system. Initial testing of AVS was followed by the development and implementation of an operational system.

As data availability and quality continues to improve, the usefulness of automated valuation systems will grow as well.

The Role of GIS in Automated Valuation

Why Spatial Data Structures?

Concepts of value is directly linked to the behavioral relationships of buyers and sellers and the attributes of the subject property. However, these relationships are complicated because real estate is a spatial commodity. Unlike other economic goods, a real estate parcel is extremely sensitive to changes in adjacent sites and the surrounding neighborhood. Urban land economists have long described this process as *land use succession.* The local real estate agent refers to this important concept with the phrase, "location, location, location."

Urban land economists note that, within the urban structure, land use classes in neighborhood clusters pass through an economic life cycle, and the land use succession process generally passes through the stages of growth, maturity, decline, and, finally, removal or renewal.

Even without delving into the particular characteristics of these general stages, the analyst should recognize that the neighborhood setting of any property is in a constant state of change. One important consideration in developing an automated valuation system with an integrated spatial (GIS) data core is to establish an analytical framework that can detect change within the succession process.

On a more practical level, the spatial data core must support the identification of similar comparables. The sales comparison approach infers the value of the subject property by identifying sales of similar properties. The spatial data core is designed to acknowledge that a buyer's decision to purchase a property is based on more than the inventory of its physical attributes. The buyer's decision includes consideration of many attributes of the subject property that extend well beyond the boundaries of the subject site, and these external attributes can best be integrated with spatial data structures.

Data Enhancement

The quality and eventual acceptance of a value estimate derived from any automated valuation system is a function of the valuation methodology applied, the presentation and believability of the resulting value estimate, and the quality of the data supporting the methodology. While most developers of automated valuation systems have focused on the integration of a limited source of data (i.e., public records and government census data), others have suggested that a multilevel data structure is necessary to support automated valuation across a broad array of property types and conditions. This is not to say that an extensive data structure is necessary to derive reasonably accurate and supportable estimates of value. However, experience is demonstrating that as users gain experience with an automated system, they attempt to apply the tools to increasingly complex valuations. As the valuation environment increases in complexity, the extent and quality of supporting data needed increase significantly.

Tables 4.2, 4.3, and 4.4 show the type of data structures most commonly associated with public record data. This information is customarily used in a tabular format and extracted from simple flat data files. Table 4.2 is a typical data table containing descriptive information about the subject property. Table 4.3 shows statistical information pertaining to the subject neighborhood, and Table 4.4 lists descriptive information on selected comparable properties.

Table 4.2

Subject Property Attribute Inventory

Subject:	1234 NE 987th St, Seattle				
Property rights appraised:	Fee simple				
Tax parcel number:	070500 0160				
Owner(s):	Orvil P. Anderson				
Based on:	Assessor information				
Property characteristics:					
Year built	1950	Bsmt. sq. ft.:	0	Garage type:	0
Bedrooms:	3	Bsmt. fin. sq. ft.:	0	Att. gar. sq. ft.:	0
Bathrooms:	1.0	Total sq. ft.:	1180	Det. gar. sq. ft.:	0
# of stories:	1.0	Lot sq. ft.:	6000	Carport sq. ft.:	0
Fireplace(s):	1	Waterfront ft.:	0	Bsmt. pkg. sq. ft.:	0
Abv. grnd. sq. ft.:	1180				
Assessed land:	66200	Previous sale date:	None found		
Assessed structure:	46700	Previous sale price:	None found		
Assessed total:	112900				

Table 4.3

Micro Neighborhood Transaction Summary

Subject area demographics (Tract: 2.00)			
Area: (square miles):	1.24	Population:	7,050
Housing units:	3,391	Ave. household size:	2.08
Median age:	37.80	Ave. household income:	$36,623
Med. home value:	$138,852		

Characteristics of the area:

Sales identified = 147	**Minimum**	**Maximum**	**Average**
Price	$103,650	$169,950	$135,896
Assessed land	52,800	80,900	66,845
Assessed structure	14,100	104,900	56,817
Assessed total	81,500	170,000	123,662
Age	9	77	42
Above ground sq. ft.	600	2,190	1,221
Bedrooms	1	6	2
Bathrooms	1.25	2.75	1.50
Lot sq. ft.	4,260	16,200	8,029

Table 4.3

Micro Neighborhood Transaction Summary (continued)

Listings identified = 29	Minimum	Maximum	Average
Price	$105,000	$295,950	$151,651
Assessed land	44,900	74,200	65,179
Assessed structure	26,700	185,000	64,537
Assessed total	74,000	244,300	129,717
Age	4	63	36
Above ground sq. ft.	590	2,890	1,242
Bedrooms	1	5	2
Bathrooms	1.25	2.75	1.25
Lot sq. ft.	4,260	12,488	7,974

Table 4.4

Comparable Property Attribute Spreadsheet

Sale 1:	12801 25th Ave. NE, Seattle	S/W 0.24 miles
Sale 2:	13045 23rd Pl. NE, Seattle	S/W 0.20 miles
Sale 3:	14032 20th Ave. NE, Seattle	N/W 0.43 miles
Listing 1:	13325 19th Ave. NE, Seattle	S/W 0.37 miles
Listing 2:	1228 NE Brockman Pl., Seattle	S/W 0.67 miles

Attributes	Subject	Sale 1	Sale 2	Sale 3	Listing 1	Listing 2
Selling price	$0	$136,000	$127,500	$125,000	$127,000	$189,950
Assessed	$112,900	$123,100	$127,500	$130,600	$113,500	$188,900
Selling date	95.42	94.56	94.03	94.64	95.34	95.33
Living area	1,180 sq. ft.	1,300 sq. ft.	1,330 sq. ft.	1,420 sq. ft.	910 sq. ft.	1,740 sq. ft.
# stories	1	1	1	1	1	2
Abv. grd. sq. ft.	1,180	1,300	1,330	1,420	910	1,740
Bsmt. sq. ft.	0	0	0	0	0	0
Bsmt. fin. sq. ft.	0	0	0	0	0	0
# bedrooms	3	3	3	3	2	3
# bathrooms	1	1	1	1.75	1	2.5
Lot sq. ft.	6,000	6,190	8,179	9,600	9,500	8,234
Age	45	42	42	42	23	7
# fireplaces	1	1	1	1	1	1
Att. garage	0	360	200	0	290	430
Det. garage	0	0	0	0	0	0
Bsmt. garage	0	0	0	0	0	0
Carport	0	0	0	250	0	0
Covered pkg.	0	1	1	1	1	1
Waterfront ft.	0	0	0	0	0	0
Basement grade	0	0	0	0	0	0
Basement type	0	0	0	0	0	0
Building grade	100	100	100	100	100	110
Building cond.	110	110	110	110	110	110
Sewer type	1	1	1	1	1	1
Topography	1	1	1	1	1	0
View (any)	0	0	0	0	0	0
View city	0	0	0	0	0	0
View lake	0	0	0	0	0	0
View lake/river	0	0	0	0	0	0
View mountain	0	0	0	0	0	0
Net adjustment	$0	$-9,299	$-3,709	$-2,907	$-2,257	$-63,497
Adjusted price	$125,000	$126,701	$123,791	$122,093	$124,743	$126,453
Estimate of value:	$125,000					

Figures 4.16 and 4.17 illustrate the unique spatial data quality that must be integrated into any automated valuation system. Because real estate is inherently a spatial commodity, the data to support real estate decision making and valuation must also be organized spatially. With data organized in a spatial context, automated valuation systems can extract significantly more useful information than could otherwise be developed from public record information. For this reason, applying the geocoding function to the data structure adds value to automated valuation systems.

Figure 4.16 is a regional display created automatically by AVS. The user simply indicates a desire to create a map by clicking a button on the summary screen. This activates a set of external macro commands, which initiates activation of the SPANS Map program. The user does not need to learn the mapping program in order to benefit from the program's use.

After the regional map is created, the analyst uses the mouse to identify the extent of the macro neighborhood area and to generate a macro neighborhood display (Figure 4.17). While the extent of the data could be automatically determined by identifying the minimum and maximum coordinate values of the data points, experience has shown that mapping is more useful because it enables the user to identify easily the boundaries of the area to be displayed. The subject property's proximity to schools, water service, census boundaries, and school districts are only a few of the factors considered by the user in determining the extent of the neighborhood map. Again, the analyst uses the mouse to define the boundary of the neighborhood, external macros are activated, and the map is created automatically.

A micro neighborhood display (Figure 4.18) is then automatically created by AVS using the same neighborhood boundaries specified by the user in the previous mapping step. The map locates many important point and vector features, such as schools, parks, shopping centers, highways, and railroads. In addition, this map identifies the locations of the comparable sales and listings used to price the subject property; their locations are labeled with numeric codes from 1 to x, with x identifying the last comparable used to price the subject. The number "1" identifies the

Figure 4.16
Regional Display

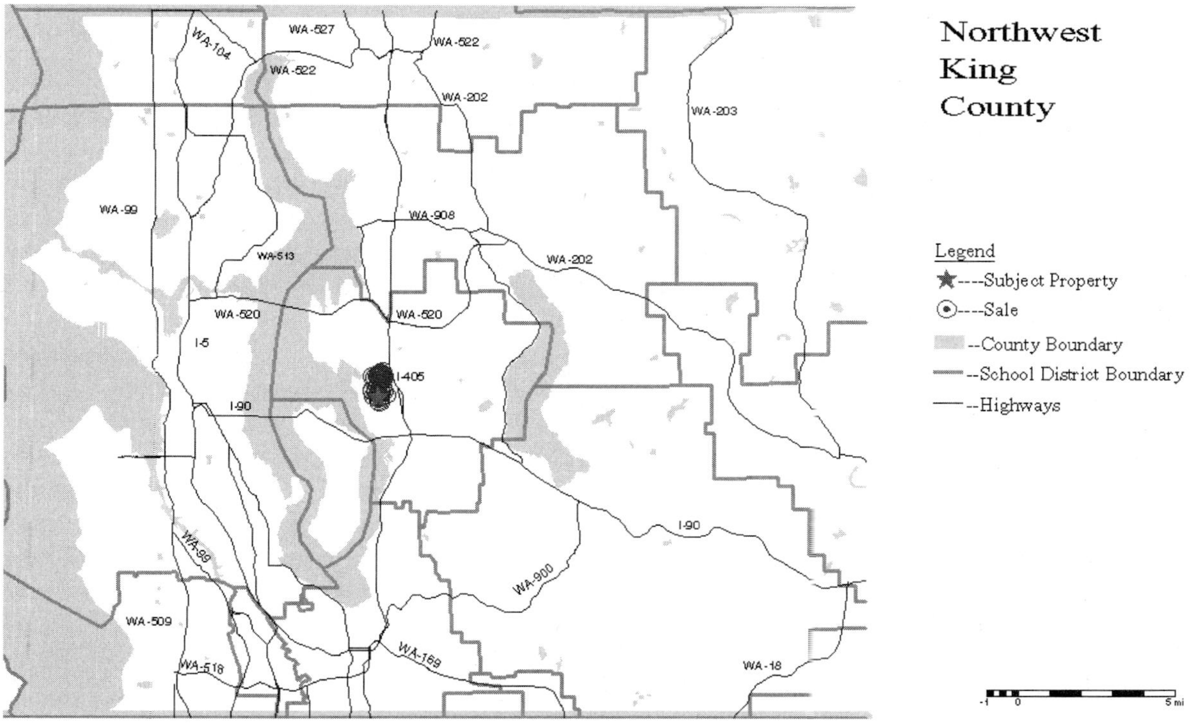

Figure 4.17
Macro Neighborhood Display

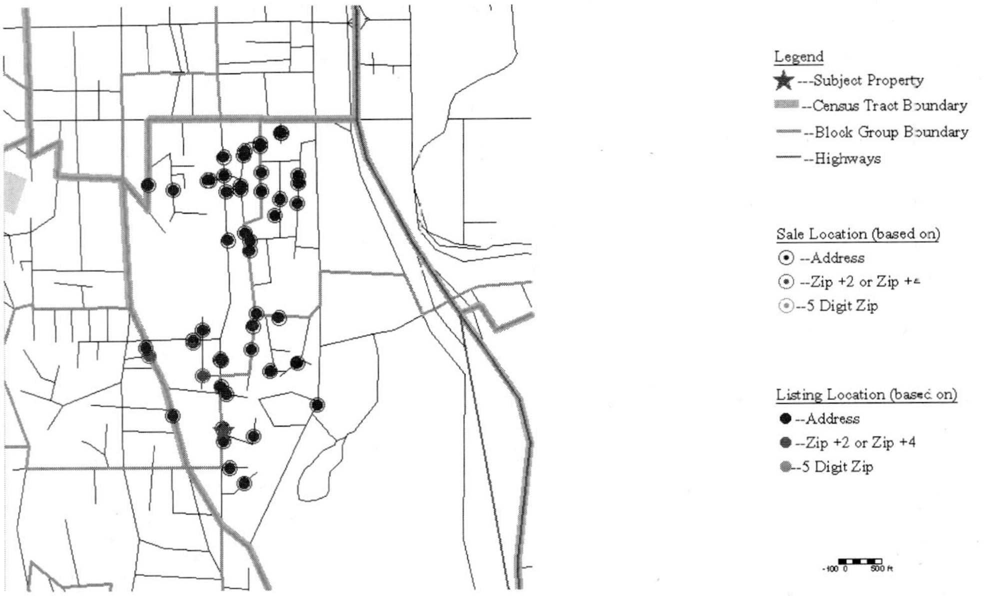

A color copy of this figure can be found on page 191.

Figure 4.18
Micro Neighborhood Display

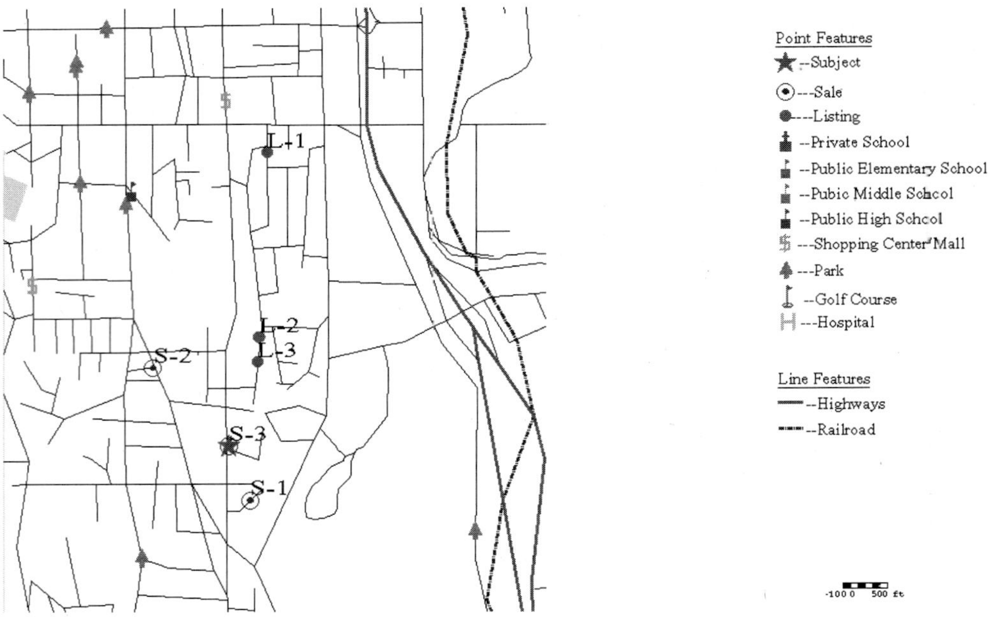

A color copy of this figure can be found on page 191.

location of the comparable or listing determined to be most similar to the subject and the numbers "2," "3," "4," through "x" identify the locations of the remaining comparables and listings ranked in terms of similarity. An "S" before the code number indicates a comparable sale and an "L" refers to a current listing. This ranking is all accomplished by AVS with no user intervention.

Note that Figure 4.18 integrates various types of GIS line and point data. The line data indicate highways, railroads, power lines, local streets, etc., while the point data identify value-influencing features such as the locations of schools, parks, hospitals, and shopping centers. Because home buyers use this information in their pricing decisions, it is important to integrate and simultaneously display these two types of GIS data. As discussed previously, vector data for the entire country are available in TIGER files. Companies like Geographic Data Technologies, Inc. (GDT) and Business Locations Research (BLR) take this base layer of data and enhance, expand, and resell it. The AVS product has integrated GDT data into its spatial structure.

CONCLUSION

By combining GIS spatial data structures with property inventory attributes, a far richer data set is made available to the automated valuation process. This richer data set is better able to capture the attribute trade-offs being made by the market and therefore can contribute to more acceptable and believable estimates of value.

The goal of an appraisal is to reflect market value from the viewpoint of the market, stripped clean of any personal bias by either the appraiser or an individual market participant. The integration of spatial data structures into automated valuation systems facilitates a better understanding of typical market behavior and provides appraisers with an efficient tool to help focus and refine their professional skill and judgment.

chapter 5

Application of GIS to Real Estate Appraisal Problems

by Bruce R. Weber, MAI

My first experiments with computer mapping began in the mid-1980s, with what was described as a "visual database" program. The first application was the creation of a mapping system that could be used to analyze business conditions and forecast the demand for real estate. Some of the methodology was described in an article published in *The Appraisal Journal* in 1990.[1]

That early article summarized my interest in business cycles and noted the need for appraisers to develop forecasting abilities. It went on to discuss how tools typically used in the stock market can be applied to quantify demand in the real estate market. More specifically, it noted that data on employment recorded in the Standard Industrial Classification (SIC) can be added to the visual database so that econometric modeling can be performed.

A 1995 publication, *The Handbook of Real Estate Portfolio Management,* contains a number of chapters which recommend the analysis of very small markets based on the analysis of employment by SIC. In this text some of the most highly regarded real estate advisors in the United States describe what went wrong in the real estate market in the late 1980s and how problems can be prevented in the future.

The examples in this book show basic GIS operations. Additional examples of the use of GIS for other types of real estate analysis may be found by typing http://www.MarketForesight.com into your Internet browser.

The editor of the handbook, Joseph L. Pagliari, Jr., CFA, states, "The well-documented collapse of commercial real estate values in the late 1980s and early 1990s caused many of us involved in the serious issue of real estate investment to carefully re-examine our investment paradigms and methodologies." He goes on to note, "The fundamental analyst/investor is considered to be more likely to uncover discrepancies between price and value."[2]

The difference between price and value is stressed. Price is the amount for which one can buy a property, but it is not necessarily the same as value. For example, the price of real estate can be driven up by uninformed buyers acting in a frenzy, but its value is determined by its future income stream, which is best quantified through market analysis.

1. Bruce R. Weber, MAI, "Application of Geographic Information Systems to Real Estate Market Analysis and Appraisal," *The Appraisal Journal* (January 1990).
2. Joseph L. Pagliari, Jr., CFA, *The Handbook of Real Estate Portfolio Management* (Burr Ridge, IL: Irwin, 1995), xvi.

The Handbook of Real Estate Portfolio Management goes on to note that the 1980s were characterized by subjective judgments. Investment decisions were "based on a real estate professional's experience, network of contacts, and gut instincts. In the 1990s, growing institutionalization has increasingly demanded that even the most subjective decisions have to be backed up with quantitative support." Also, "Demographic changes and employment growth must be monitored to determine the supply and demand dynamics of individual markets."[3]

The market to be analyzed, termed the *micro market,* or *economic location,* is identified as the area in which demand for a property originates, which is often a five to seven mile distance from the property. GIS is an excellent tool for monitoring changes in such an area.[4]

The two examples provided in my article presented data from the Phoenix, Arizona, and San Diego, California, office markets, which suggested that both markets were becoming oversupplied by 1987. Few clients could be convinced that serious problems were starting to occur in the Southern California and Phoenix business cycles, and I subsequently made a change from working as an independent appraiser to employment as a review appraiser. I started an appraisal review department at an agency to resolve differences in opinions of value that appraisers had reported for problem properties owned by failed banks.

CHAPTER OVERVIEW

Many of the methods described in this chapter were first implemented in an attempt to reconcile widely differing opinions of value. Most of the variation in the value opinions resulted from vastly different projections of demand for real estate. One appraiser formed the opinion that a half-leased office building would be 95% occupied in three years, with rents increasing by 4% per year, while another appraiser forecast stable occupancy and stable rents—for the same building at the same point in time. These appraisers made subjective judgments (like Pagliari describes) and provided no quantitative support.

Resolving such discrepancies provided an excellent opportunity to experiment with methods to improve the appraisal process. Since the properties were often sold within a year of the dates of appraisal, it was easy to determine which appraisal techniques were the most accurate in predicting selling prices. It has been five to seven years since the sales, so the overall accuracy of the forecasts used by the appraisers can also be noted.

This chapter is divided into the following sections:

- The use of GIS in appraisal review
- The use of GIS in business cycle analysis and forecasting
- How GIS can help appraisers respond to criticism

The examples are grouped by property type, and the discussion of each will help illustrate the various types of spatial analysis that GIS can perform. Although some of the methods employed may not have practical applicability in appraisal, the examples are included to demonstrate basic GIS concepts. They were also chosen to highlight the most valuable role of GIS: its ability to convey efficiently considerable amounts of information on spatial relationships that are not discernible in tabular data and to integrate disparate databases.[5]

The examples in the first section show how GIS was implemented as part of the appraisal review process. The maps have been changed and street names have been omitted for purposes of confidentiality. In one instance a map of a different area, not the area where the subject property is located, is shown here, but the underlying data are accurate and relevant for purposes of the case study.

The examples presented in the second section relate to a consulting assignment and show how GIS can be used to expand the services that an appraiser can provide. The purpose of the

3. Ibid., 99.
4. Ibid., 289.
5. Gilbert H. Castle, III, ed. "Profiting from a Geographic Information System," *GIS World Books* (Fort Collins, CO: GIS World Inc., 1993).

study was to help the manager of a real estate fund decide which assets to sell and which to buy to improve the returns from his portfolio.

The manager of the fund knew the price he could get for the property, which is the purpose of a conventional appraisal. However, he did not have a good feel for what each asset would likely be worth in three to five years, so he was unsure as to which assets to sell and which to buy. An analogy of common stocks can be used to help focus on the objective of the assignment. Anybody can tell what an actively traded common stock will sell for today. The real challenge is in being able to predict what it is likely to be worth in three to five years.

The third section was written because many people believe that appraisers are not capable of forecasting supply and demand for real estate. If this were true and the definition of value is the present worth of future benefits, appraisers would have to change their practice or the definition of value. Without the ability to forecast supply and demand, the only nonresidential property that appraisers would be qualified to appraise would be real estate subject to a long-term lease to a creditworthy lessee, with a known income stream as described in the lease.

The most powerful aspect of GIS for real estate analysis is its usefulness as a tool to quantify and forecast the demand for a given property within its most relevant market area. This is the essence of value.

USE OF GIS IN APPRAISAL REVIEW

Example 1. Dot Density Mapping: Basic Quantification of Demand

A review of appraisals relating to a special-purpose recreational property provides a good example of the basic functionality of a GIS. The subject property had experienced problems since it was built. The management had changed frequently and the operation did not seem to be profitable.

One of the appraisers provided what is commonly known as a "ring study," which summarized the demographics of the population within a 10-mile radius of the site. (Some appraisers do not even consider how many people are required within a market area for a property to be feasible.) Other appraisers did consider the area population and thought that it was adequate to support the recreational center. Income and expense statements were not available, so the appraisers simply assumed (another case of subjective judgments) that the property had suffered from consistently poor management since the time it was built.

One type of GIS mapping that a review appraiser can use in such a situation is thematic mapping, in which shading or dots are used to represent values. The map in Figure 5.1 shows where the population resides within a 10-mile radius of the subject and its competition, Properties A and B. The population is represented by dots, with each dot indicating the location of 50 residents.

The figure was created by mapping out the population by zip code boundaries. Using this map, a review appraiser can communicate quite rapidly that almost all of the population within the 10-mile radius is located at the extreme southeasterly corner of the ring seven to 10 miles distant from the subject property. Thus, the concentration of people who were thought to provide the demand for the subject recreational facility live much closer to the competitive centers. The subject property is likely to have a poor capture rate because it is too far from the majority of the population.

Point thematic maps can also be used to show the locations of important linkages that may create demand for the property. Certain types of businesses can be represented by color-coded dots, or dots of various sizes can be used to indicate the locations of major employers that may have an effect on the demand for a recreational facility. The thematic mapping of the incomes in the area can also provide useful insight into variations in demographics, which may explain the lackluster performance of what appears to be a well-designed real estate venture.

Analysis of the population density and incomes of the area residents may be critical in the valuation of recreational/special-purpose properties. Rent comparables are generally not available for such properties. When they are, the rents paid probably reflect the demand for the comparable property based on its competition and the density of the population surrounding the comparable property, not the

Figure 5.1
Use of GIS to Find Population Density

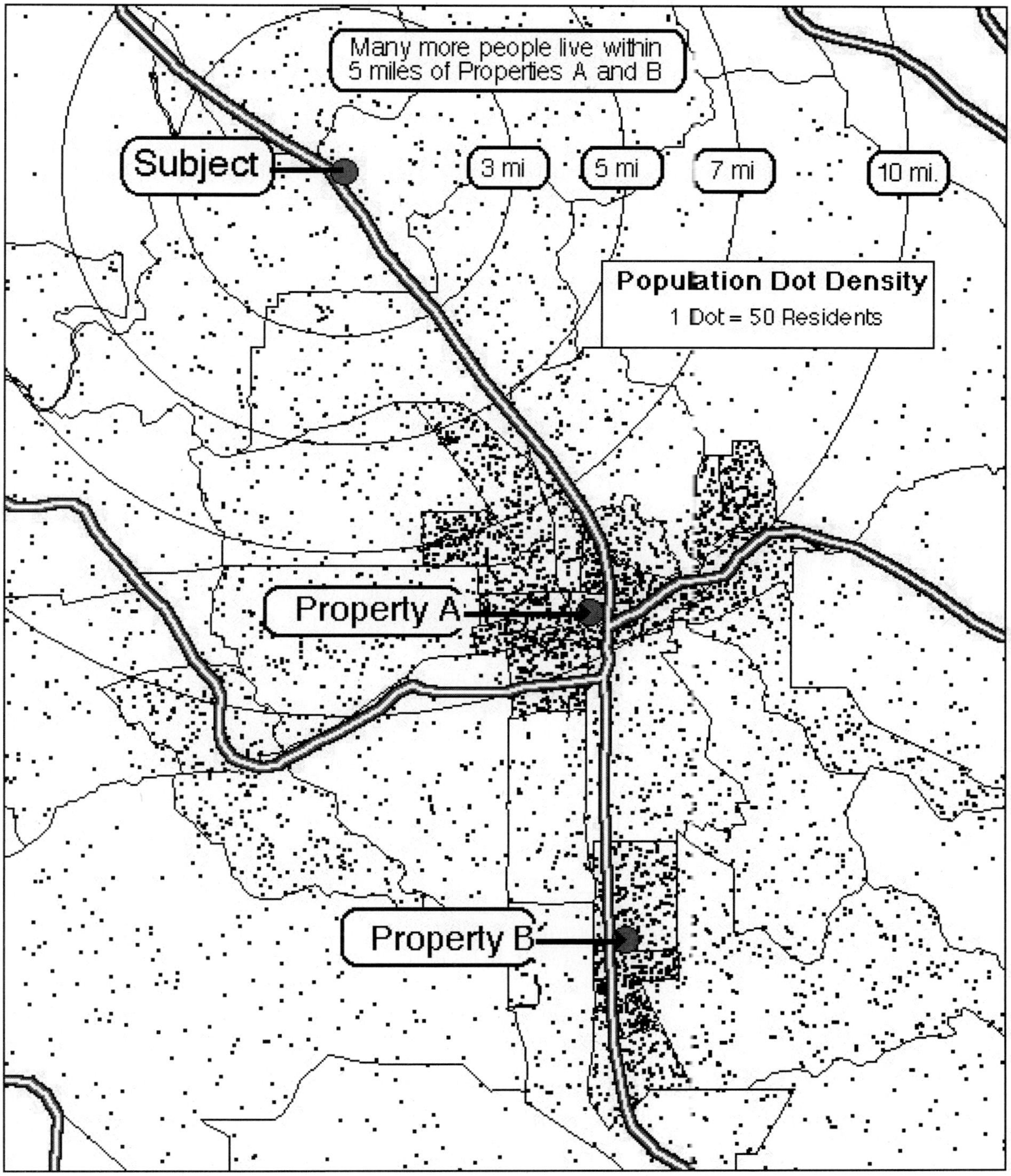

likely rent for the subject. The sales comparison approach is not useful for the same reasons. Application of the cost approach would require an analysis of economic demand to quantify external obsolescence. Thus, the estimated sales volume for the subject may be the best starting point.

Similar problems may be encountered in the valuation of a hotel property. Business costs must be subtracted from the sales revenues to determine if there is enough income to pay for management, furniture, fixtures, equipment, and reserves before finding out if there is an adequate return to the real estate.

The sales potential from the most likely trade area can be estimated with a GIS using the Thiessen Polygons or drive-time models. Geographic information systems use spreadsheet-like formats to store data for each attribute of a unit of geography such as a zip code area. One spreadsheet might have the population for each zip code area, while another has estimates of average disposable income. The GIS can multiply the population from one of the spreadsheets by the income from another and then multiply the results by the analyst's estimate of the average percentage of income spent on the type of activity the property provides. The result is the sales potential for the property.

These calculations can be done by hand, but few review appraisers have the time to accumulate the data and perform the mathematics for a market area that comprises, say, hundreds of census block groups. To add another refinement, the analyst can calculate the potential income spent at the subject property by residents living in the bands located one to three, three to five, five to seven, and seven to 10 miles from the subject. These figures can be multiplied by decreasing capture rates for each of the bands. Such an analysis would definitely be too time-consuming for a review appraiser who does not use a GIS.

Example 2. Calculation of Areas and Volumes: Contaminated Soils

Another very basic task that can be performed using a GIS is measurement of areas. This example concerns a property that has contaminated soil. The extent of contamination must be identified to describe the environmental problem and its likely impact on value. In this type of analysis, the appraiser should obtain an expert opinion from a health risk assessor regarding the health risk associated with the contamination to help quantify the financial risk involved.

The site in question has been contaminated by a leaking underground storage tank. A plot plan of the site is shown in Figure 5.2. Soil borings have been used to measure the concentration of contaminants at five-foot intervals at various locations on the site. The contour lines shown on the map connect areas of equal contaminant concentration at each depth.

In the site plan shown in the figure, contamination was investigated to a depth of 30 feet. The consultant can plot out these contours and easily calculate the area between each of the contour lines. Multiplying the area between the contour lines by the depth indicates the volume of soil that has been contaminated; the concentration of the contaminant reveals the volume of the contaminant in the soil.

Some water quality control boards provide tables that can be used by property owners and others without technical expertise.[6] These tables can be consulted to determine whether or not a specific contaminant must be remediated based on the types of soils, the distance between the contaminant and the water supply, and the concentration of the contaminant. This information can be used to evaluate the level of financial risk that may result from the contamination and the likelihood that it will have to be remediated.

Example 3. Use of Point-In Polygoning to Find Comparables

An appraiser can use GIS to search for properties that are located a certain distance from a given point or points. Concern about contaminated real estate has become widespread in recent years, and some property owners have contended that their land has suffered a decline in value simply because it is located near a known source of underground contamination. One way to help prove or disprove this allegation is to find other properties located near contaminated sites that have been sold.

6. Bruce R. Weber, MAI, "Stigma - Unquantified Risk?" a paper presented at the 1996 Cutting Edge Conference of the Royal Institution of Chartered Surveyors in Bristol, England.

Figure 5.2
Use of Overlays to Calculate Irregular Areas

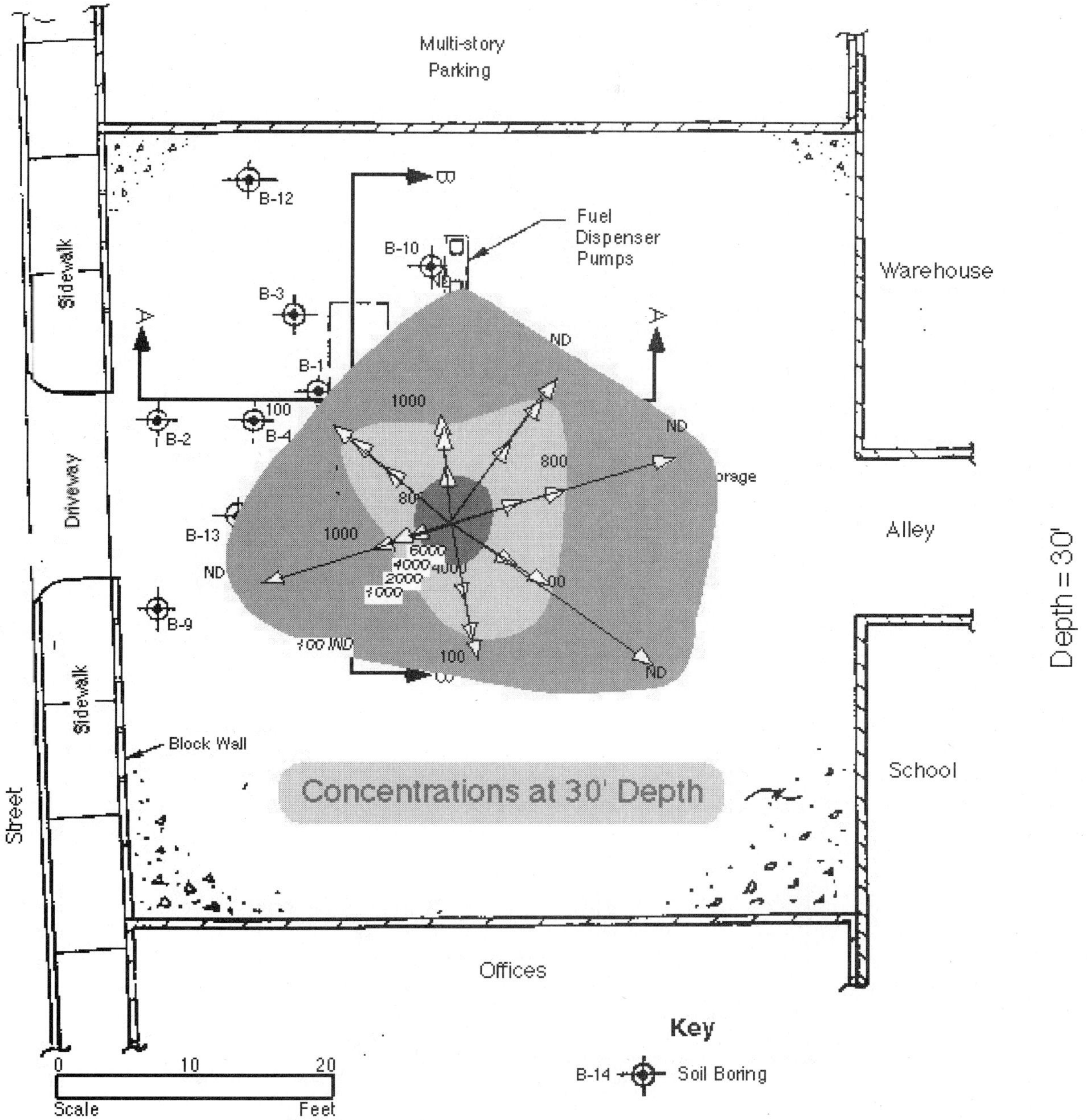

The points shown on the map in Figure 5.3 indicate the locations of parcels of real estate that have sold with some form of contamination. Many systems allow users to create buffer areas around such points, lines, or irregular areas on an electronic map and then search for sites located within the buffer area. In this case a circular buffer area with a 1,000-ft. radius was created around each contaminated site. If the appraiser has land sales that have been geocoded, he or she can perform a spatial query with the GIS database to search for parcels of land within the buffer areas that have been sold.

The inset map (Figure 5.4) provides additional detail. An appraiser who understands the environmental risk resulting from various types of soil contamination can then confirm the sales with

Figure 5.3
Use of Point-in Polygoning to Find Comparables

Figure 5.4
Inset Map with Added Detail

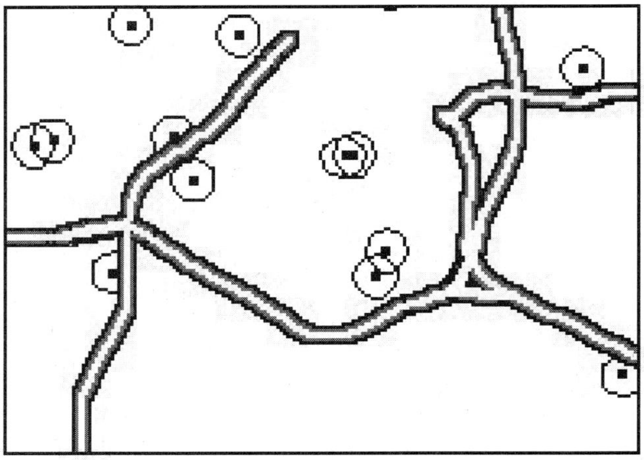

the buyers and find out which ones are the most relevant to the site being appraised. A site contaminated with immobile heavy metals, for example, may be less likely to cause stigma than a site contaminated with a more pervasive hazard. The appraiser can then discern what effect, if any, nearby contamination has on the parcel prices.

Example 4. Calculating Overlapping Areas: Biologically Valuable Land

The previous examples have illustrated the basic operations of presentation, thematic mapping, distance measurement, and area calculations which are typically available on a GIS. The remaining examples will relate to demand, supply, and equilibrium data—critical information that can be categorized and analyzed through the use of a GIS.[7]

Use of GIS for the Supply Side

San Diego has some of the greatest concentrations of biologically valuable land in the country. The Multiple Species Conservation Plan (MSCP) is a new, focused effort to protect large amounts of biologically valuable land that are connected by *biological linkages.*

Biological linkages are required for endangered species to travel from habitat in one area to habitat in another. Large, contiguous blocks of biologically valuable land in certain areas have been designated for conservation. When biologically valuable land is lost to development in an area, however, mitigation land must be acquired in a nearby preserve area and maintained into perpetuity to compensate for the loss of habitat elsewhere.

The MSCP makes use of *gap analysis.* If the goal is to preserve, say, 1,000 acres in one biological submarket, and the Bureau of Land Management (BLM) already owns 750 acres in the area which are being preserved, the gap analysis would show a net demand for 250 more acres in the area. An analysis of the supply and demand for mitigation land within the biological submarket can then be made. This analysis is important because, once the goals for preservation within a biological submarket have been met, there will no longer be any demand for mitigation land in this area.

Some systems have the ability to import what is known as a *raster image* of a map. This means that a map can be electronically scanned and its image can be saved on a computer disk. This image can then be imported into the GIS and, as shown in Figure 5.5, be used as a base map that is overlaid with data. In this case, sales of acreage have been placed over the base map.

Base maps are created by local planning departments using a very sophisticated GIS. The use of raster maps allows images generated by these systems to be used, but precludes the need for data compatibility that could add considerable complexity to the analysis. It is not always easy to import data from one GIS into another.

The base map in Figure 5.5 has been shaded to show the value of the habitat throughout the MSCP. Sales of acreage and biological values can be visually correlated to identify comparables that are likely to have similar biological values. A similar methodology was applied to Figure 5.3 to search for property sales near to sources of contamination. This procedure is referred to as *categorizing data.*

The locations of the acreage that has been sold are geocoded and the sales are mapped as overlays to the base map. The GIS is then used to categorize the sales by, say, price and to color-code the data.

Tracing an outline with GIS to calculate the area of high value habitat The map in Figure 5.6 shows the boundaries and acreage of lands that have a very high habitat value and the locations of the linkages between these lands. It was created by tracing the outlines of lands with very high habitat value onto the imported raster map. Once the areas have been traced over the base map, the base map can be made invisible and the land area of all of the outlined areas can be calculated with one command.

7. Castle, 97.

Figure 5.5

Overlay Map of Sales by Price per Acre

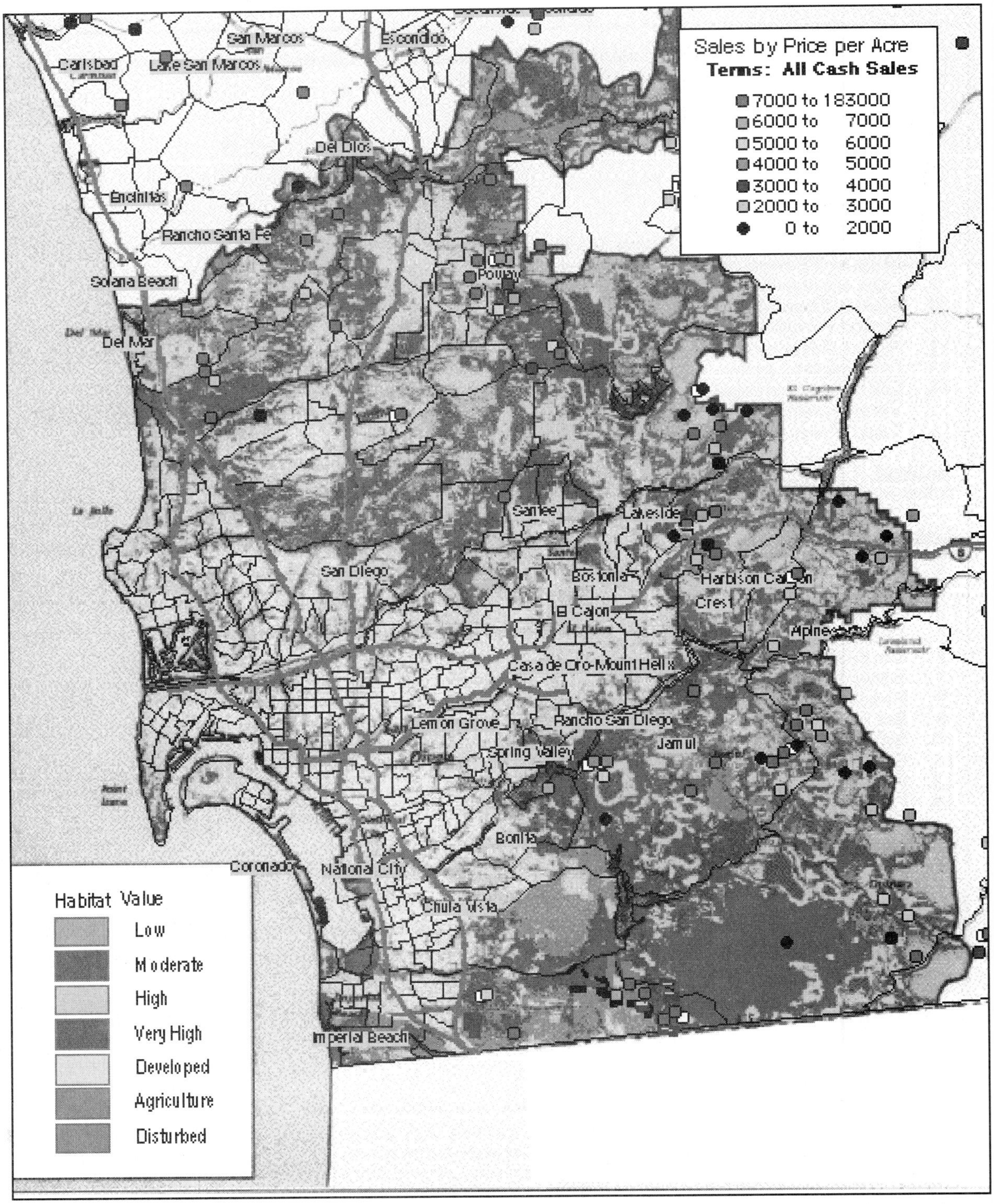

A color copy of this figure can be found on page 192.

Figure 5.6
Supply and Demand by Biological Submarket

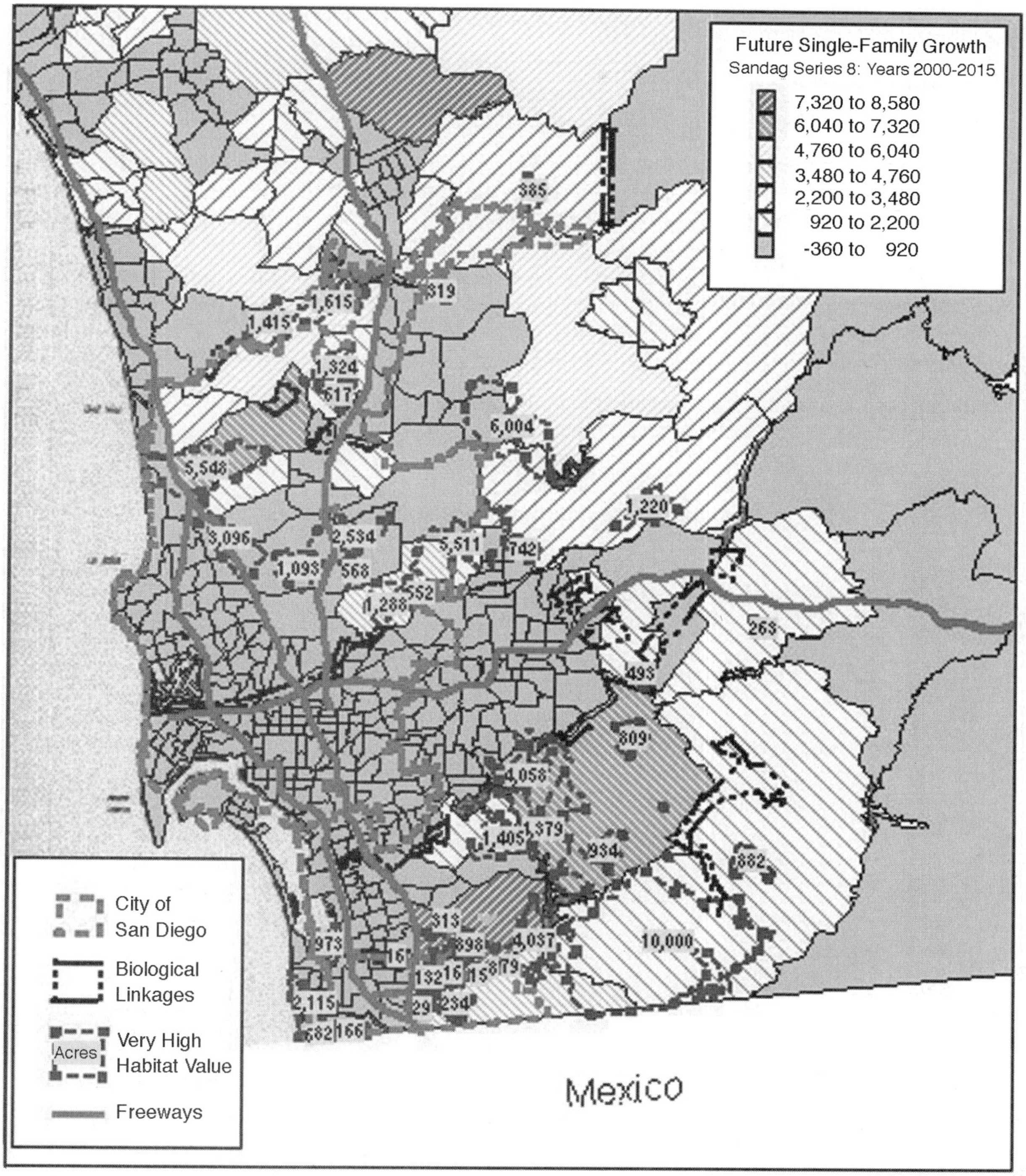

A color copy of this figure can be found on page 193.

The base map that was in the background has been replaced with a map of census tracts, which has been shaded to reflect the relative population growth forecast for this part of the county. The red areas are anticipated to be high growth areas. These submarkets are likely to generate a much greater demand for mitigation land than the low-growth submarkets. (The growth scale has not been shown to avoid cluttering the exhibit.)

Maps such as these can be used to communicate the relative supply and demand conditions for mitigation land by submarket. A much more specific analysis of supply and demand can be performed using a GIS that is capable of overlay calculations.

Tracing an outline of the portion already preserved Figure 5.7 shows a blow-up of some of the biologically valuable habitat in this submarket identified by the red boundary. It also shows the boundary of the BLM land that already has preserved status. The supply of mitigation land in this area consists of the land that is outside of the blue boundary that is bounded by the red boundary.

The boundary of the land preserved by the BLM was also identified by importing a scanned image of a map that shows the boundary and tracing it, which takes 10 to 15 minutes. Calculating acreages is time-consuming by other methods, but quickly accomplished with a GIS. Simply selecting the red boundary to show statistics on its image reveals that it encompasses 25,680 acres.

Calculating the net available supply Using the boundary file for the BLM land like a cookie cutter, the area outside the BLM land bounded by the high value habitat land can be calculated in a few clicks

Figure 5.7

Example of Noncontiguous/Nominal Overlays

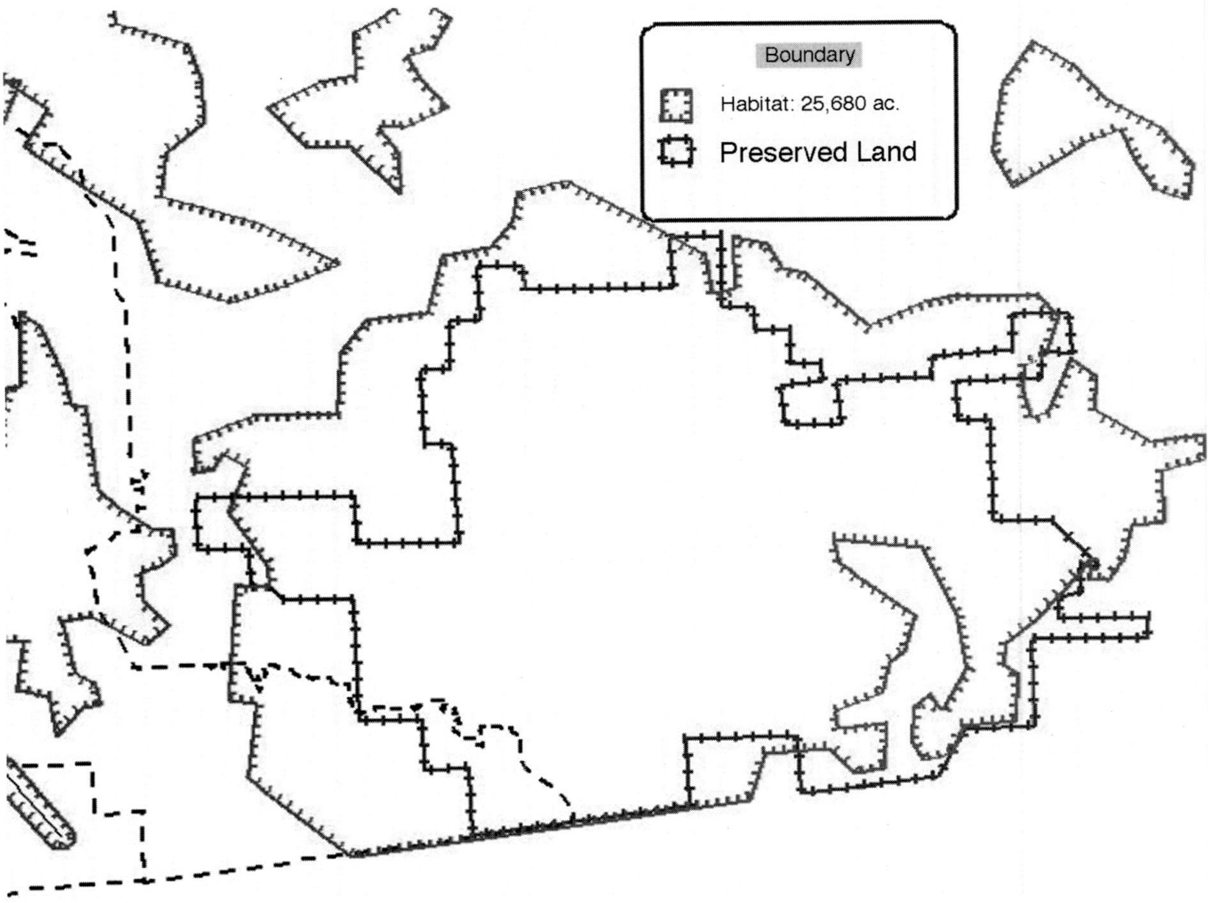

A color copy of this figure can be found on page 194.

of the mouse. Reference to Figure 5.8 shows that the net supply of mitigation land available is 8,801 acres.

Many more types of analysis can be performed to reflect the relative value of portions of a land holding. These variations can result from the mitigation ratios that may be applied if only some of the acreage is located in the preserve.

Use of GIS for the Demand Side

The demand for mitigation land can be analyzed as a derived demand. Figure 5.6 shows the locations of land that is likely to be needed for housing, while Figure 5.9, in the next section, shows the locations of job growth. The data mapped can be used to estimate the amount of land needed for development and the resultant need for mitigation land in the future. For example, forecasts of demand for land for residential or industrial use are calculated by local planning authorities. These estimates of future land use are often made by planners working with a GIS. As the preceding examples have demonstrated, appraisers can use GIS to analyze the supply and demand for any kind of real estate.[8]

Figure 5.8

Use of BLM "Cutter" to Find Supply of Land

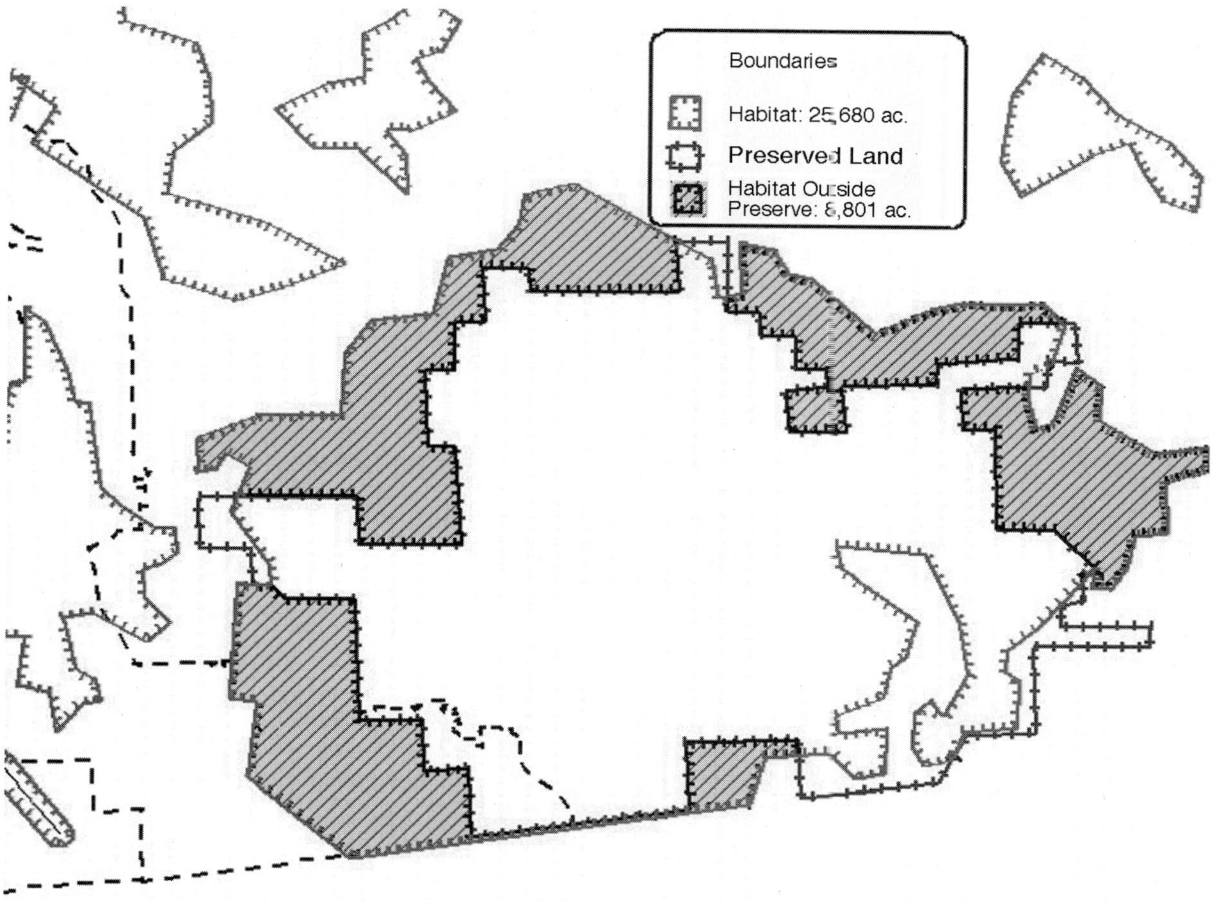

A color copy of this figure can be found on page 195.

8. For general information on real estate market analysis, see Stephen F. Fanning, MAI, Terry Grissom, MAI, and Thomas D. Pearson, MAI, *Market Analysis for Valuation Appraisals* (Chicago: Appraisal Institute, 1994).

Example 5. Use of Data Queries: An Industrial Example

Industrial demand analysis generally starts with a forecast of employment by Standard Industrial Classification, or SIC. This forecast is then translated into the square feet of industrial space demanded by each employment category. The basis for the forecast of employment is often locational quotients. With GIS, these locational quotients can be analyzed for each submarket to create small-area forecasts with the results displayed on a map.

Use of GIS for the Supply Side

Appraisers are accustomed to creating maps that show the locations of individual properties such as industrial buildings. These maps typically show the existing supply in a given submarket within a city or county.

With a GIS, the building addresses can be geocoded (by assigning longitude and latitude coordinates to building locations), mapped, and then shaded or changed in shape to reflect the results of spatial queries.[9] A large number of empty buildings shaded red, the size of a nearby retail competitor indicated for a retail study, or an important linkage shown on a map can convey useful spatial information to a client.

Use of GIS for the Demand Side

Some county planners provide forecasts of the number of employees in each census tract by major SIC category. Some of this information was downloaded via the Internet and the data was mapped (see Figure 5.9). Employment data analyzed by zip code or census tract can provide insight into the

Figure 5.9

Distribution of Employment

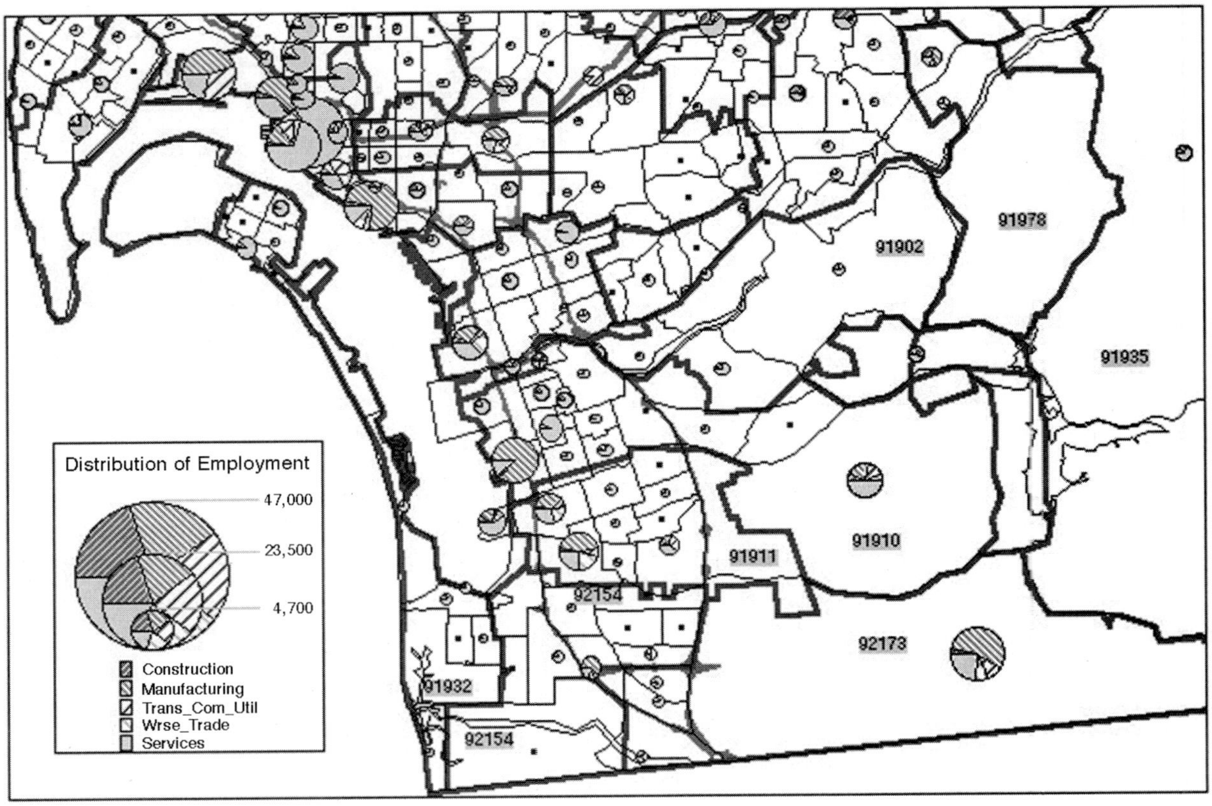

9. Weber, "Application of Geographic Information Systems to Real Estate Market Analysis and Appraisal."

economic characteristics of an area and likely changes over time. The increase in the number of industrial employees, for example, can be translated into net demand for industrial acreage.

The Use of Buffers to Aggregate Data for Comparison

Buffers, i.e., lines that are a fixed distance around an object, can be drawn with a GIS, as illustrated in Figure 5.3. The object, which is known as a *polygon,* can be straight, curved, or highly irregular. Once the buffer has been created, it can be used as a selection tool to group data that are within the geographical area of the buffer area.

For example, Figure 5.10 shows a buffer around the U.S./Mexican border that was created to compare percentage changes in manufacturing employment within the buffer area over time to changes in manufacturing employment in the rest of the border states. The map was created while reviewing a market study to check the forecast of employment prepared by the appraiser.

The buffer area included all the land area located five miles from the border. Manufacturing employment within each of the zip codes within that area was totaled for the beginning of 1994. The analysis was repeated using employment data for 1996 to identify the trend in manufacturing employment in the border area.

After the data were aggregated by the GIS and totaled, the trend in total manufacturing employment within the border states was also calculated. This analysis helped the reviewer demonstrate to the appraiser that the market study was flawed and based on assumptions that were proven to be unreliable.

Figure 5.10

Change in Manufacturing Employment by Zip Code

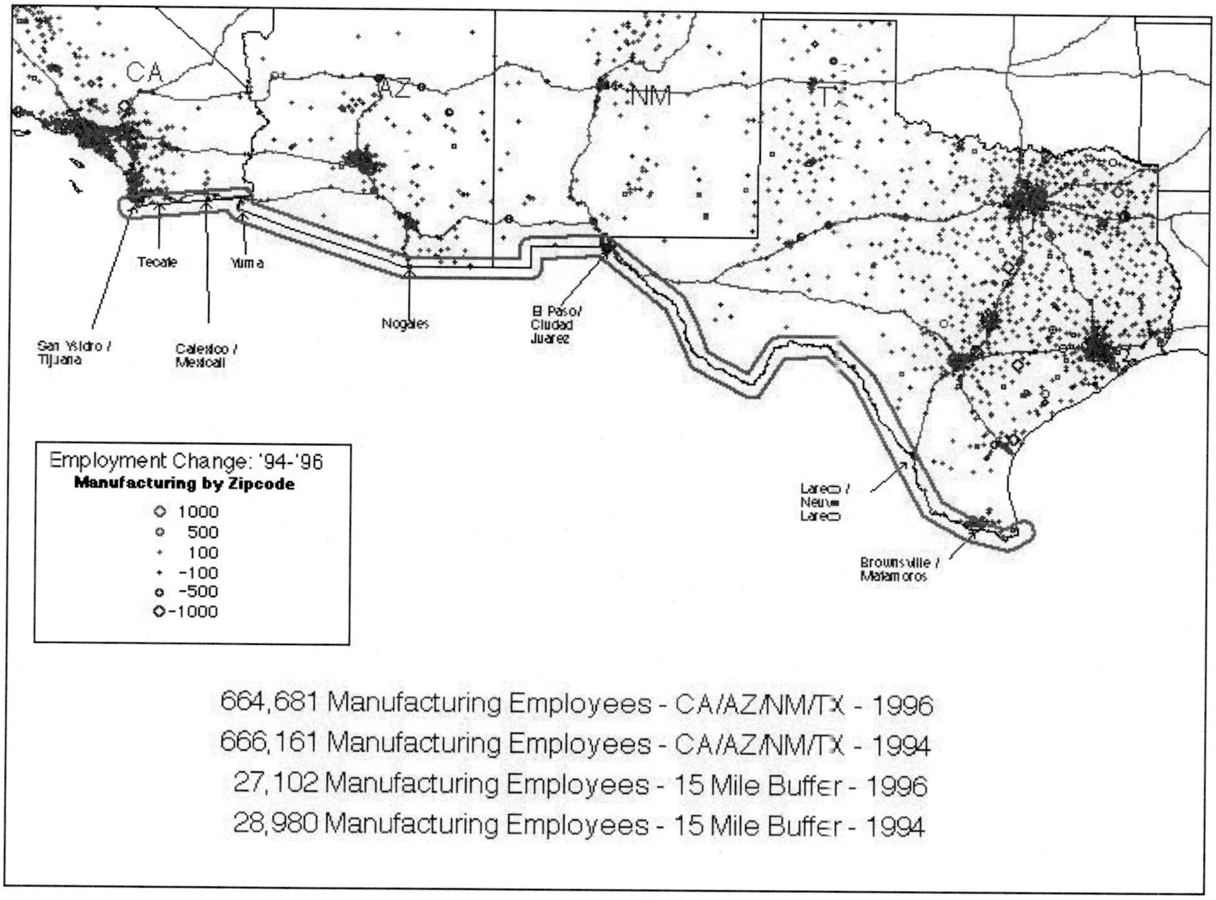

Use of GIS in Retail Analysis

GIS may well prove most valuable as a tool for retail analysis. Various types of mapping systems have been created in the past in an attempt to choose sites that will be successful store locations.[10] GIS can be used to create such maps much more quickly.

Using the basic principles of market research developed 30 to 40 years ago, GIS can be used to devise much more powerful retail gravity models. These models facilitate rapid analysis of the effect of new competition on existing store locations.

Example 6. Large Retail Site

Use of GIS for the Supply Side

The map shown in Figure 5.11 was used as part of a review of two market studies performed for a large retail site. The Uniform Standards of Professional Appraisal Practice (USPAP) do not specify the extent of review that is expected when a review appraiser is retained to review market studies. My preference is to prepare an independent supply and demand analysis so I can get a feel for the market.

The competitive supply of retail space can be retrieved, geocoded, and mapped for the market area being studied. I prefer to show the locations of competitive centers as dots of graduated sizes, reflecting the size of each center. The locations of the regional malls in the area are shaded.

Figure 5.11

Aggregating Sales for Trade Areas A and B

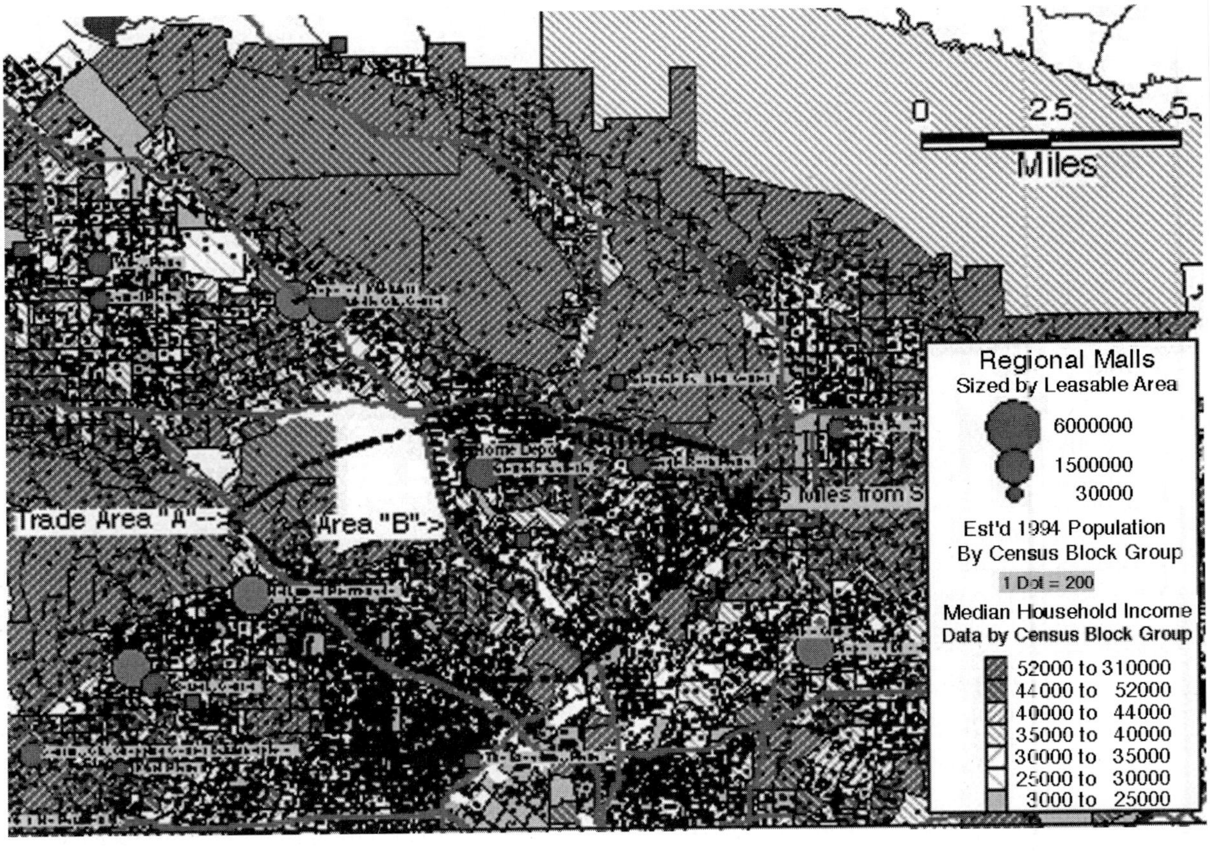

A color copy of this figure can be found on page 196.

10. William Applebaum, *Guide to Store Location Research* (Menlo Park: Addison-Westley, 1968).

The locations of proposed shopping centers are also geocoded and indicated on the map with red circles. A "what if?" analysis is then done to calculate how the development of each of the centers would affect the sales volume and resultant economic rent of a retail property in Trade Area B. For example, Trade Area A might contract toward Trade Area B as a result of the development of a proposed shopping center at the westerly boundary of Trade Area A. This addition to supply could reduce the sales volume and economic rent of stores in Trade Area B.

Map of supply by SIC Another interesting analysis can be performed by plotting out the locations of the competitive retailers by SIC. The background of the map can then be shaded to indicate the typical per capita expenditures by product type for each of the income groups in the market area. Important linkages can then be examined on the map and sales volumes can be estimated for each of the competitive stores.

The creation of analogous situations for demand comparison (analogs) GIS has also been used to analyze the trade areas surrounding acreage that has been purchased by a major retailer such as Wal-Mart. Once the locations of the land sales have been plotted on a map that overlays demographic statistics, it is easy to calculate how many people typically reside within three- and five-mile radii of the site and what their total purchasing power is. This type of analysis can be used to show that a particular retail chain would be one of the most likely buyers of a property and that the buyer may be inclined to pay as much for the subject as for a comparable if the demographics and competitive supply are similar.

Use of GIS for the Demand Side

GIS is also exceptionally good for analyzing demand. Once a system is set up, any type of boundary can be drawn on the map and used as a selection tool. Using the boundary, the appraiser can find out how many people are in the area, what their incomes are, and, depending on the data included with the system, how much they are typically going to spend on a wide number of retail purchases.

As an example of the graphic power of GIS, consider a map showing a number of land sales that had occurred at various locations and reflected a wide range of prices per square foot (Figure 5.12). Application of the appraiser's conventional tools gave all the wrong answers. The smallest site with the best frontage, which would be expected to have the highest per-unit value, actually sold for the lowest price per square foot. The largest comparable with the least amount of frontage sold for the highest price per square foot. The sales all occurred within about a year of each other.

The factors that were not considered in the adjustment of land sales are differences in access and the supply and demand relationships surrounding each site. GIS can be used to calculate drive times from each of the sites to rate their access. The resulting map shows that the market area of Site C is three times as large as the market area of Site A. The analysis creates a polygon that defines the geographic area within the drive time. This polygon can be used to select information on incomes by census block group for census blocks within the polygon and to gather information on competitive centers. The data can then be used to quantify the retail sales potential of each of the sites so that a proper adjustment can be made for location. Use of GIS allows the analyst to emulate the market more closely since purchase decisions on retail sites are not based on adjustment grids that consider ordinary units of comparison such as size and shape.

The use of GIS can also help to avoid common mistakes. In regard to the recreational property described in Example 1 at the beginning of the chapter, for example, the appraiser erroneously assumed that demographic averages from a 10-mile ring study would be good enough to justify economic feasibility. Similarly, the retail map in Figure 5.11 shows extremes in income, and both exceptionally high-income people and exceptionally low-income people are unlikely to shop at mid-level price stores. The high incomes skew the average income up, so the real sales potential may be much less than the income averages indicate.

Figure 5.12
"Comparables" with Major Drive Time Population Differences

Site C has twice as many people within a 3 minute drive time as Site B and three times as many people as Site A

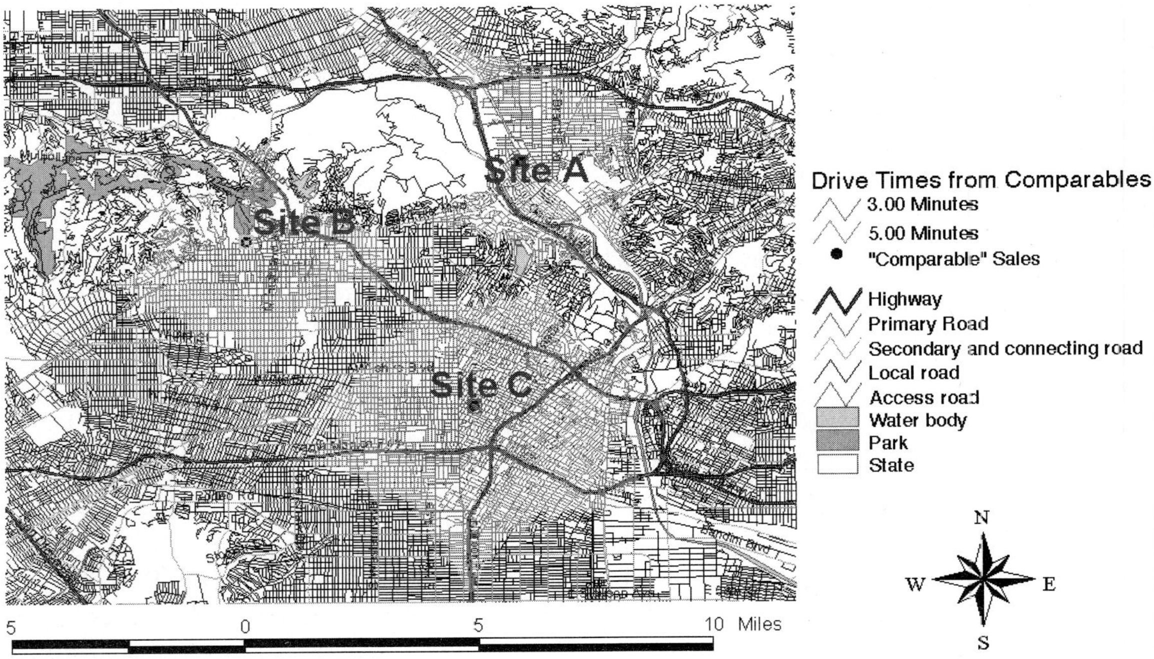

A color copy of this figure can be found on page 197.

Combining GIS Databases

Example 7. Residential Land Example

Figure 5.13 was created to answer a common problem with residential market studies: How to define the relevant market area from a demand standpoint. One method of market definition is described here and another is discussed in the subsequent section on business cycle analysis and forecasting.

The subject property is a large land holding with highly irregular topography. In Figure 5.13, data from the census are graphed using pie charts and overlaid on the base map to show the location of very high home values and their proximity to the subject. The location of the subject property is intentionally not identified.

Productivity Analysis

The productivity analysis performed suggests that the highest and best use of the subject tract is for a subdivision of custom homes. The sample map presented shows how a GIS can assist in the productivity analysis. A review of prior appraisals of the property indicates that such a land use would require an extended period of absorption. Since a multiyear forecast of demand is required for proper application of a subdivision model, an in-depth, level C market/marketability study should be requested when another appraisal is ordered.[11]

Use of GIS to Define the Market Area

One of the suggested techniques applied to define a market area is the use of census tracts. The categories for home values used in the census do not go high enough to differentiate between the expensive homes in the residential subdivisions that comprise the competitive supply in the area.

11. Highest and Best Use and Market Analysis (Course 520), developed by the Appraisal Institute, Chicago.

Figure 5.13

Using Census Data and GIS for Exploratory Analysis

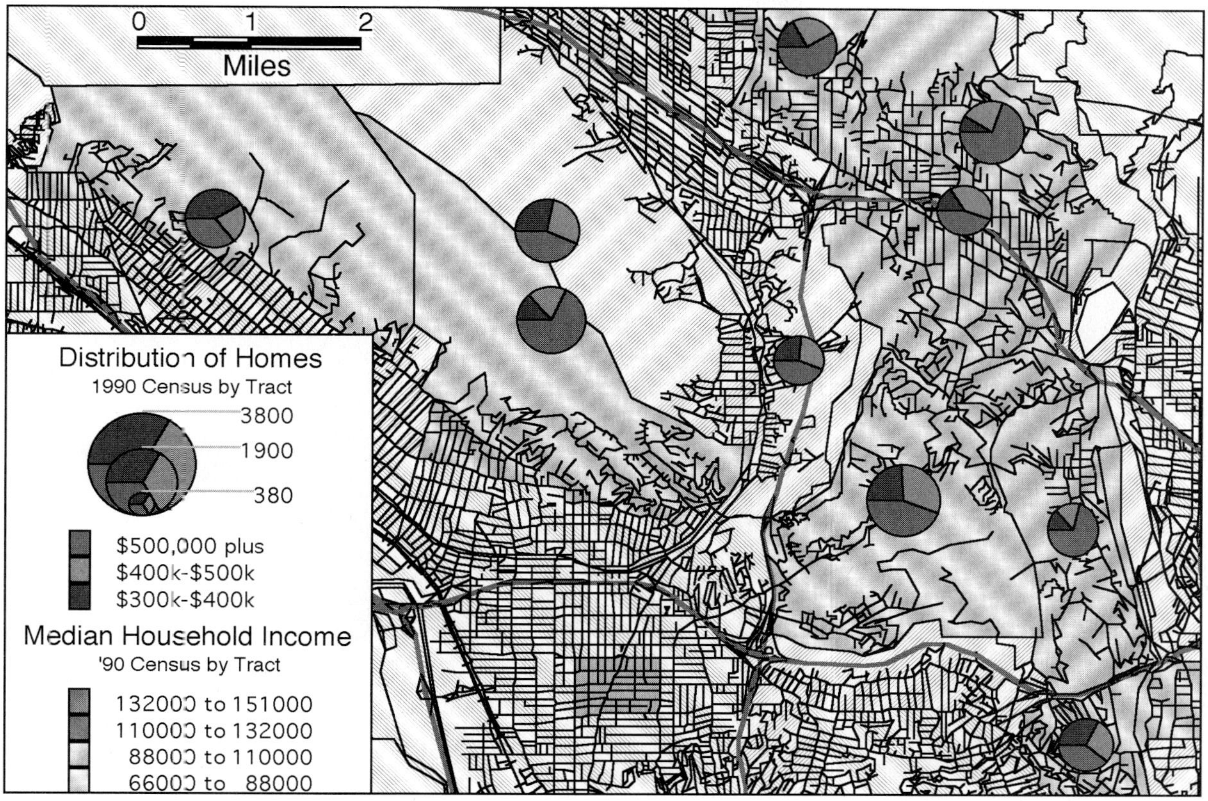

A color copy of this figure can be found on page 198.

Also, the 1990 census has become dated, so updated data on income levels in the census tract were purchased from a commercial data vendor. These data were used to prepare Figure 5.14, which shows how different high-income areas compare in income.

Figure 5.15 was created by "zooming out" from Figure 5.13. It shows the locations of the competitive supply of luxury homes over a wider area. The market can be better segmented when the spatial distribution of the supply of homes is viewed in relation to the locations of high-income employment. Experience suggests that those with high incomes who work near the subject will move up to new high-cost homes near their places of employment to reduce their commuting time.

Identifying the employers paying high salaries Figure 5.15 could be useful in delineating the primary competitive market area. Once this area has been defined by census tract, information can be obtained about the types of employment that generate the high incomes of the residents in these census tracts. Figure 5.16 shows a frequency distribution of the types of employment engaged in by residents of the high-income census tracts in the subject neighborhood at the time of the census.

Drive time data help define the market area Figure 5.17 shows a frequency distribution of the time it takes residents to get to their high-paying employment. This is useful information because the GIS also allows us to calculate market areas as a function of drive times. In this case, about 80% of the high-income population drove less than 34 minutes to their places of employment. Add-on modules for GIS are available to perform drive-time analysis. The module can be used to highlight the streets that are within a 34-minute drive time of the site being analyzed (see Figure 5.18).

Figure 5.14
Updated Demographics Differentiate High Incomes

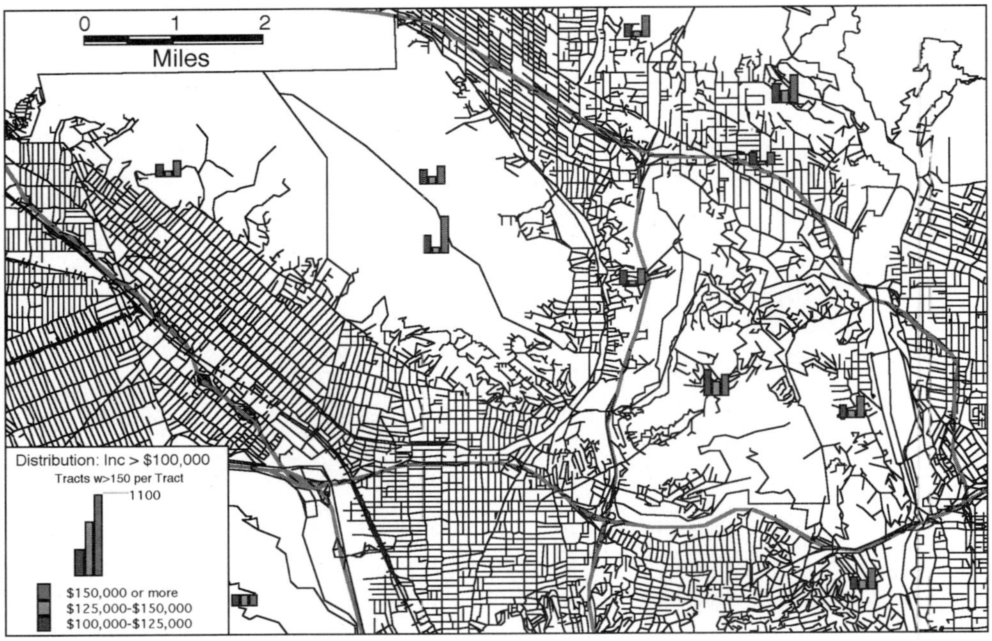

A color copy of this figure can be found on page 199.

Figure 5.15
An Income Database to Locate Competitive Areas

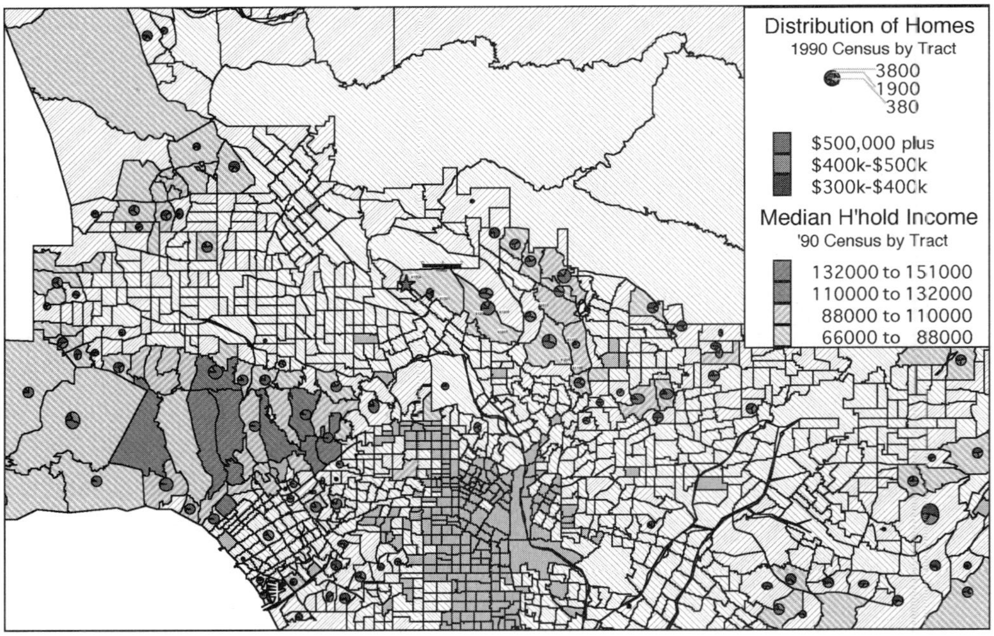

A color copy of this figure can be found on page 200.

Figure 5.16
Frequency Distribution of Employment

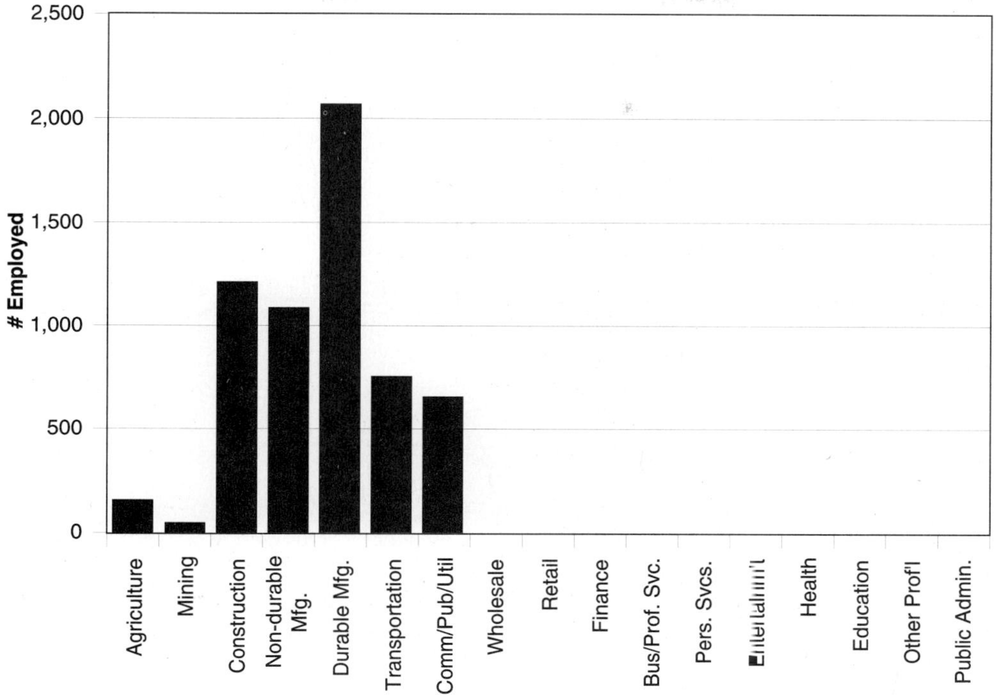

Figure 5.17
Frequency Distribution of Drive Times from Tracts to Employment

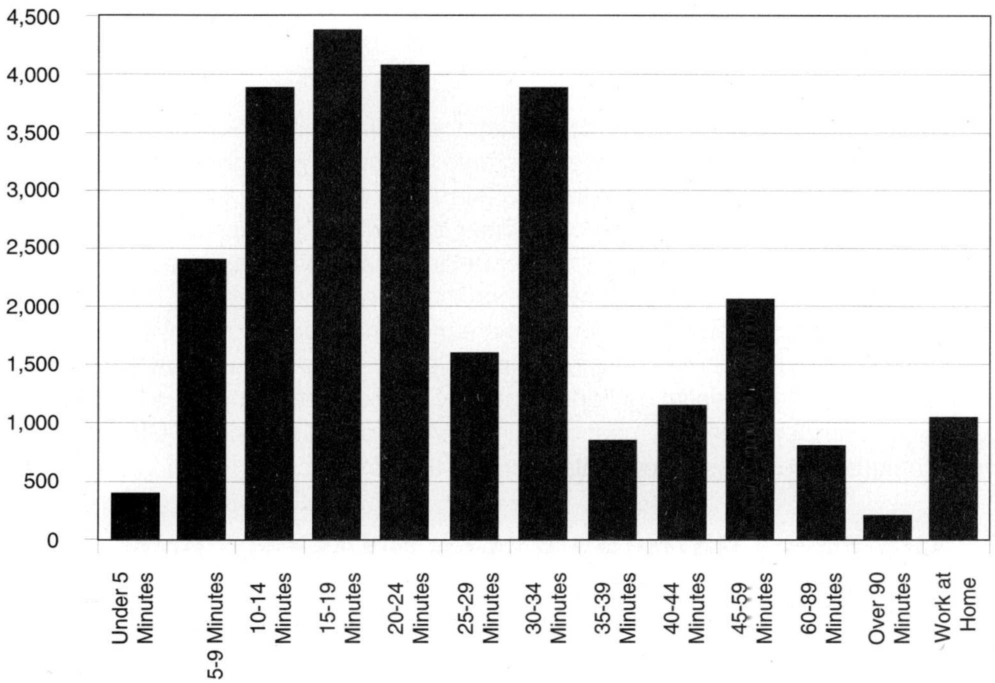

Figure 5.18
Streets Within a 34-Minute Drive Time of Subject Site

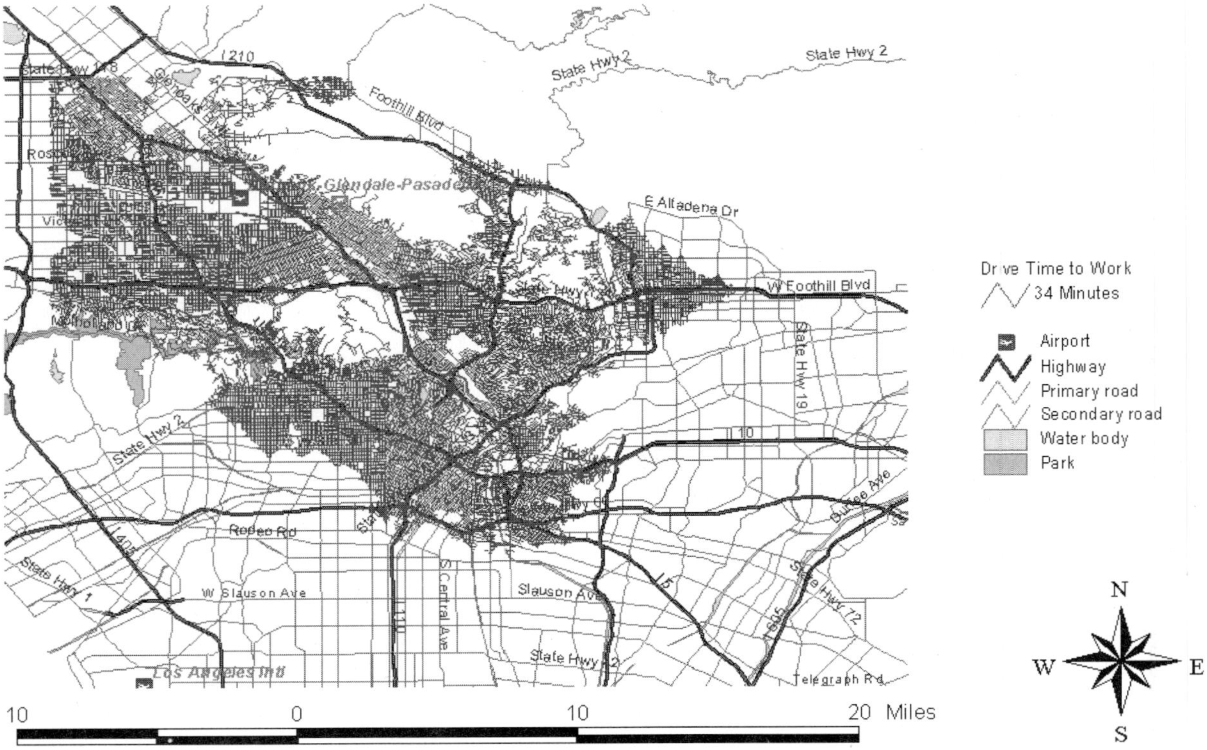

A color copy of this figure can be found on page 201.

Locating databases on employment Major employers can be readily found for any given market area. In this case high-income employment in the competitive tracts was concentrated in the SIC categories of transportation, communication and utilities, construction, and manufacturing. The locations of the major employers in these categories were mapped with different icons in Figure 5.19, identifying the employers by SIC.

Final definition of the relevant market area To present the final definition of the subject's market area, the analyst uses a large number of databases to create GIS overlays. Figure 5.19 shows the partitioning of places of employment that are within the 34-minute drive time of two competitive sources of existing supply—one to the northwest and the other to the west. Trends in employment within these core industries, which are located within the dashed line on the map, are likely to account for much of the demand for custom homes to be built on the property in question.

Locational quotients for the area delineated are then analyzed to identify the prospects for changes in demand for high-income housing within the competitive market area. Smaller businesses can also create increased residential demand. One way to take smaller businesses into consideration is to map out changes in employment over time by SIC category using a GIS. The use of forecasting techniques is discussed in greater detail in the next section.

USE OF GIS FOR BUSINESS CYCLE ANALYSIS AND FORECASTING

In addition to the use of GIS for appraisal review and market analysis, GIS can be used for many conventional consulting assignments undertaken by fee appraisers. The first example in this section concerns the overbuilt office market of Phoenix, which was mentioned at the beginning of the

Figure 5.19

Using Disparate Databases to Define a Market Area

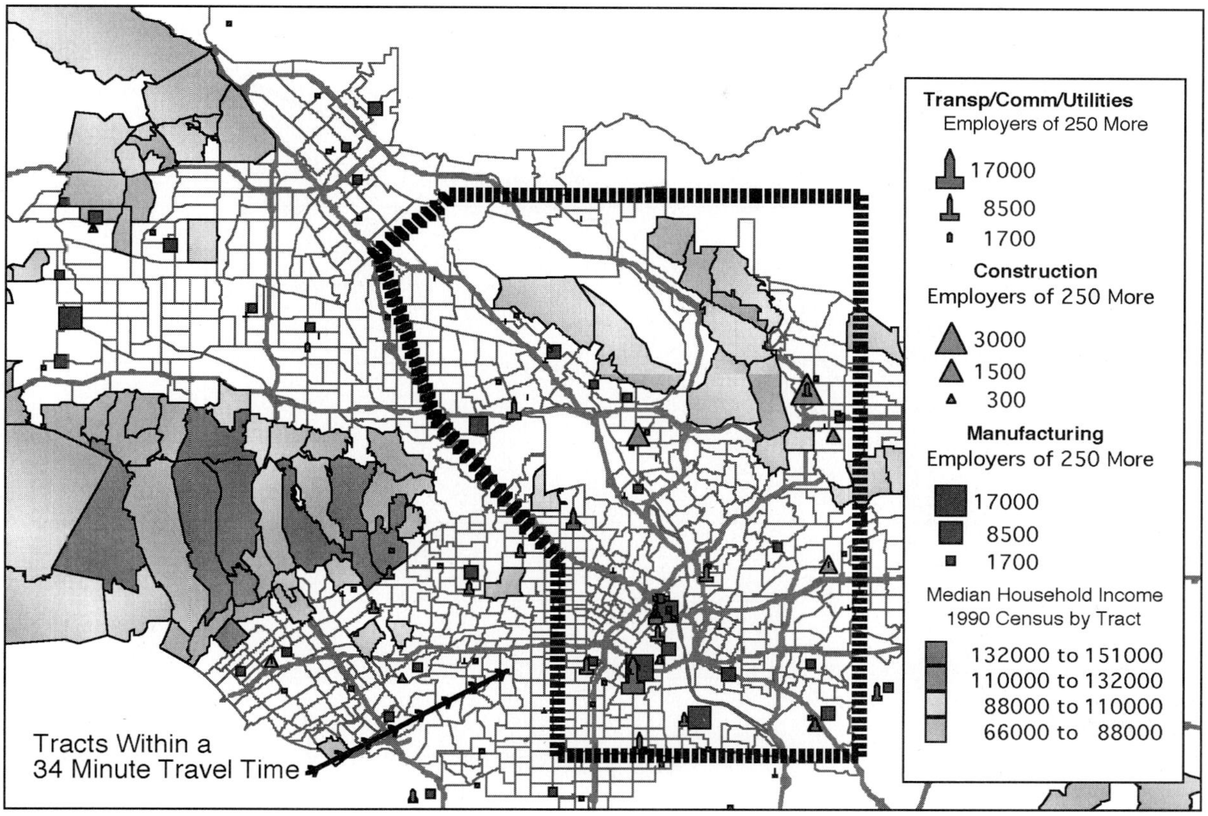

A color copy of this figure can be found on page 202.

chapter. The scenario described illustrates problems that resulted from some appraisal methodologies that were commonly used in the late 1980s.

The Human Model of Forecasting

It seems to be human nature to base expectations of the future on the assumed continuation of recent historical trends. This tendency is illustrated in the following example.

Example 8. The Phoenix Office Market

Figure 5.20 graphs the trend in total employment for the Phoenix/Mesa metropolitan statistical area (MSA). The boom in employment through 1986 and 1987 suggested a continuing need for new office space. This assumption seemed logical given the reasonably consistent relationship between employment and the demand for office space.

Office absorption during 1986 provided a reasonably good estimate of the likely absorption rate in 1987. During 1988 and 1989, however, use of historical rates of employment growth and absorption resulted in a major oversupply of office space. The situation became even more ominous in 1990 when total employment declined.

Real Estate Market Forecasts

The Appraisal Institute has published a series of research reports dealing with the problem of forecasting real estate markets. A 1987 research report, *Real Estate Appraisal and Market Analysis,* summarizes the proceedings of a research forum on the topic. One of the participants stressed the

Figure 5.20

Use of Historic Absorption to Predict Future Office Demand—Phoenix/Mesa MSA

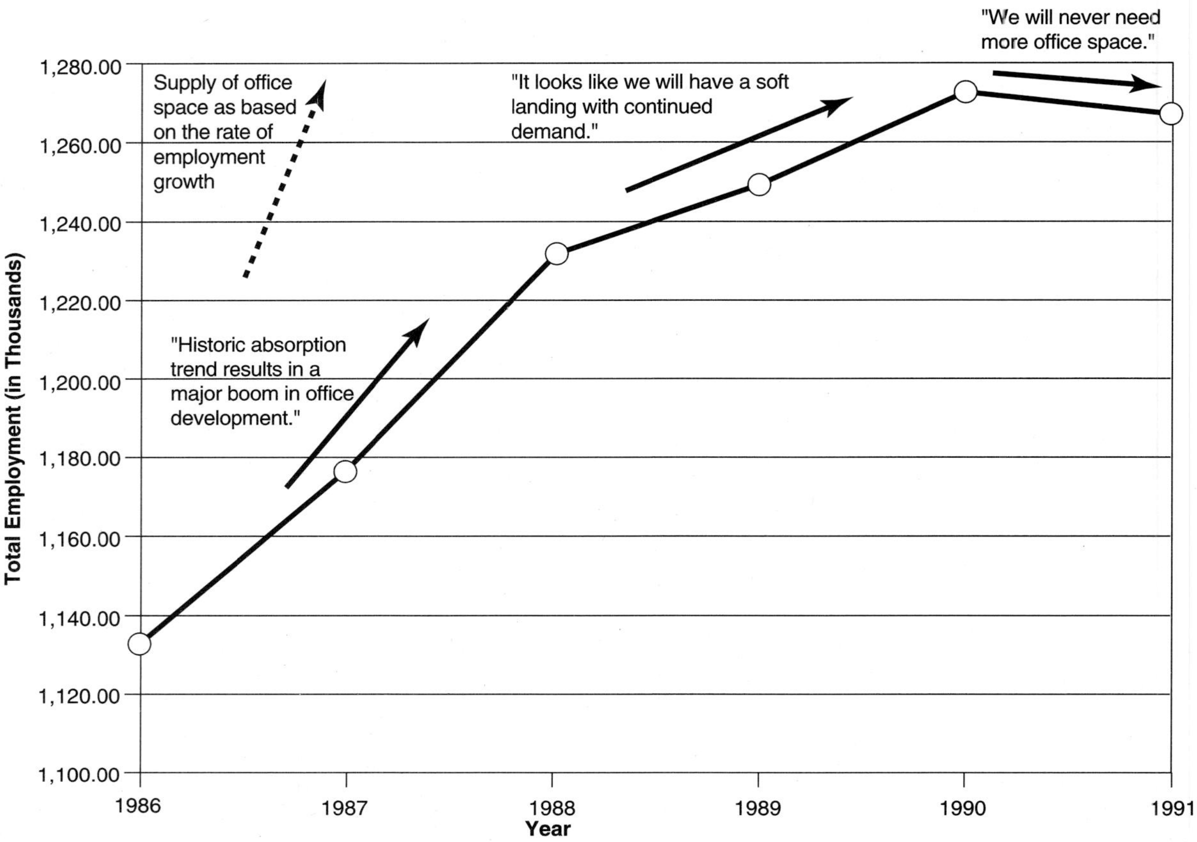

need for five-year forecasts of office demand. Joseph Rabianski, PhD, another participant, who teaches real estate at Georgia State University, noted that the analyst must make efforts to obtain estimates of basic employment. He explained that it is people outside the local economy who create the demand for goods and services that causes the economy to grow.

Forecasting: Market Determinants Affecting Cash Flows and Reversion, a research report published in 1989, presents a model for forecasting the demand for office space. The model starts with the definition of the relevant market area and then calls for an economic base analysis, which includes evaluation of current and future employment, population, and income trends.

Alfred Gobar, PhD, an economist who has been studying real estate markets for more than 30 years, wrote a portion of the Appraisal Institute's 1993 publication *Real Estate Market Analysis: Supply and Demand Factors.* Dr. Gobar has examined the relationship between employment at the local level and the demand for various types of real estate in an effort to forecast real estate demand. In his discussion of the office market, Dr. Gobar notes the importance of distinguishing between employment that services the local community and employment that functions as an export industry.

The example that follows incorporates the suggestions of Rabianski and Gobar, forecasting future employment by basic employers located within the most relevant "economic area" of a large apartment complex as defined by GIS.

Example 9. Defining the Relevant Market Area for a Large Apartment Complex

Figure 5.21 shows a map of employment sites for residents of an apartment complex in western Riverside County, California. The address of each employer was geocoded, mapped and identified with a star on the map.

Figure 5.21
Delineation of Primary and Secondary Market Areas

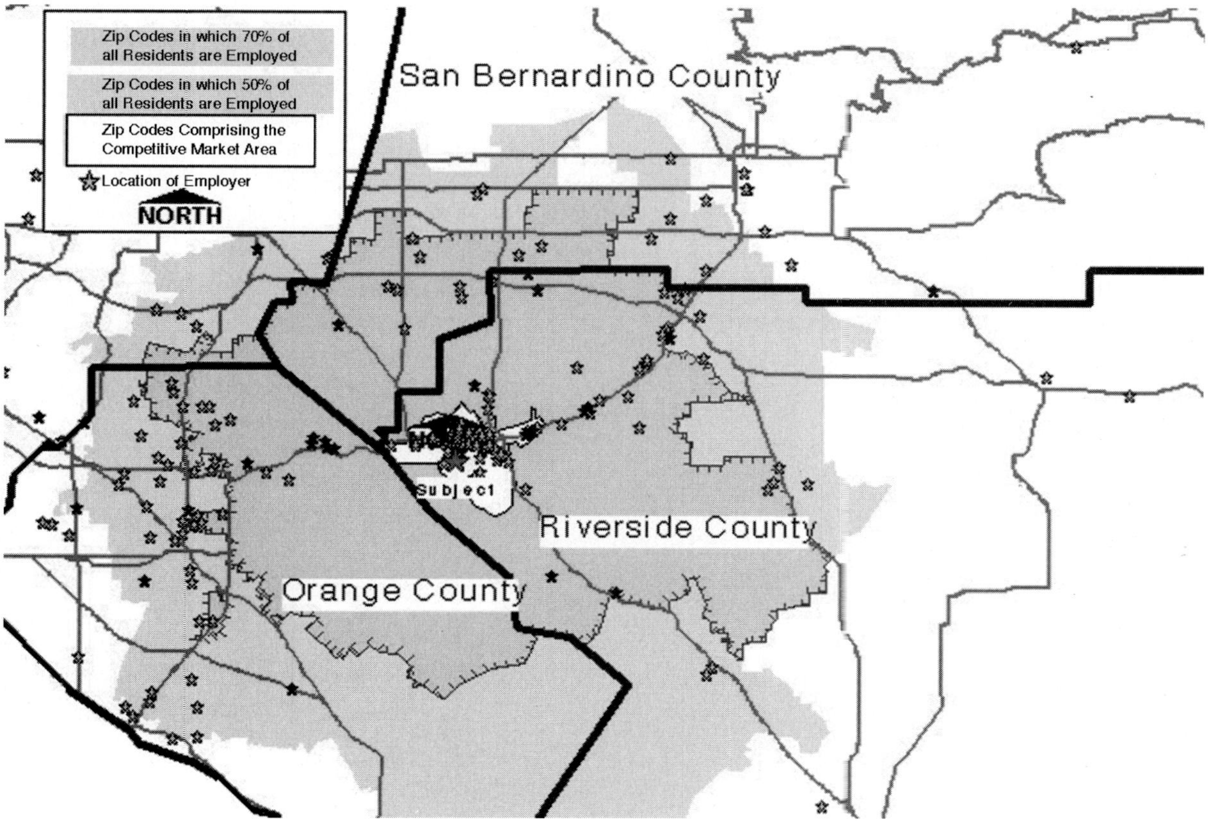

A color copy of this figure can be found on page 203.

A convex hull was then used as a boundary around the locations of half of the employers that are located closest to the apartment complex. This boundary identifies the primary trade area. A second boundary was generated to contain the 50% market area plus the next 20% of the employers closest to the subject. This boundary identifies the secondary market area. The GIS software was then used to select the zip codes with centroids located within the primary market area (the area shaded in yellow) to identify the geographic area for which employment is analyzed.

The analysis used zip code boundaries rather than census tract or block group boundaries because employment data are available by zip code. Figure 5.22 shows the results of mapping changes in employment by zip code over the last two years. Note that demand is not homogeneous throughout a given county. The employment changes that account for apartment demand occur in a three-county area. The changes in employment also suggest that different capitalization rates may be appropriate for different parts of a given county.

All of the employment located within the zip code areas comprising the primary market area can be categorized into basic and nonbasic employment and sorted into industrial groupings for which three- to five-year forecasts are available. The forecasts can then be used to estimate the amount of basic employment within the submarket; the nonbasic employment can be estimated by applying employment multipliers. If these estimates are made annually, the changing business cycle can be noted by the declining rate of change in employment revealed in the annual forecasts of basic employment.

This methodology was used to forecast the change in employment for the competitive market area shown in white at the center of the map in Figure 5.22. Employment was forecast to increase by 1,791 from the beginning of 1997 to the beginning of 1998. The actual increase was subsequently

Figure 5.22
Mapping Changes in Demand

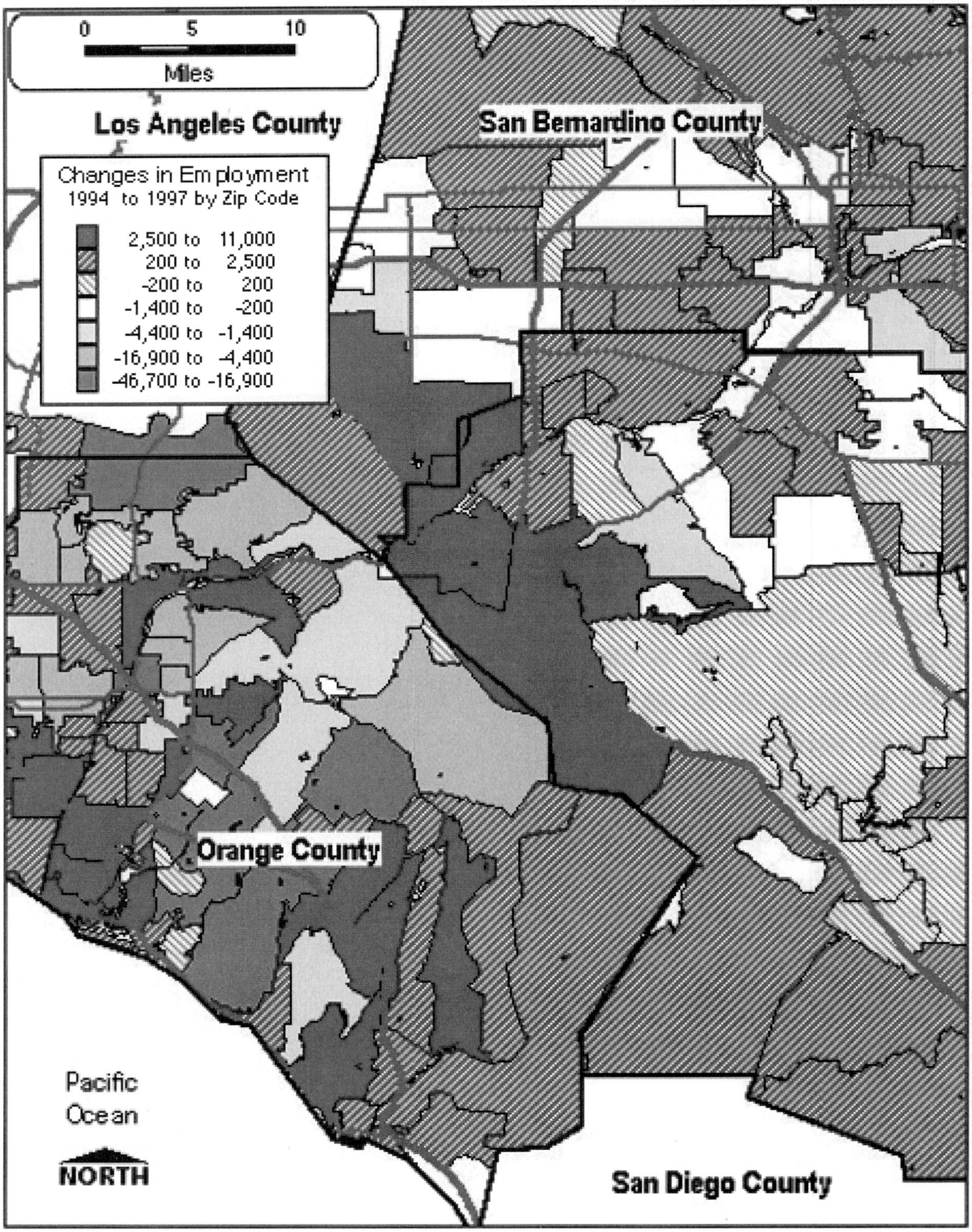

A color copy of this figure can be found on page 204.

estimated to be ¯,664. Rents were forecast to increase by 5%, while the actual increase was 4.6%. Both of these forecasts were within 7% of the actual figures, a good result considering that the rents remained stable for the five to six years since the apartments were built

HOW GIS CAN HELP APPRAISERS RESPOND TO CRITICISM

Appraisers have been under fire. Many of the papers presented at the first meeting of the International Real Estate Society held in Stockholm criticized appraisers for allowing rampant overbuilding to occur. Real estate researchers from Australia, Great Britain, Sweden, and many other parts of the world noted that financial difficulties have resulted from the rampant development of real estate during the late 1980s. Some of the papers presented are quoted below:

> Recent criticisms of valuers in Australia (Property Economic Task Force) in their use of DCFs highlights the problem of using constant compounding growth assumptions in DCF cash flow forecasts. Evidence shows that, historically, cash flows and market capitalization rates affecting reversionary values are cyclical. To assume that cash flows will continually increase, particularly at a compounding rate, subject to fixed rental review terms and ratchet clauses, etc. over the investment period is likely to be unrealistic.[12]
>
> We need do no more than look around Central London to see that property cycles matter to all of us. The massive capital, labour and other resource investments in constructing buildings which stand idle is a testament to the profound and lasting impact of ill-timed development and ill-informed urban planning and management. The property cycle has many and often damaging effects on key sectors of the economy.[13]
>
> Traditional real estate valuation models stabilize cash flow variables in a single or multi-year pro forma, assume efficient markets, and impute property return/risk expectations into a current overall market capitalization rate to determine value. This traditional valuation framework is biased towards trend analysis and often assumes constant annual changes in rents and expenses and constant terminal valuation capitalization rates over a seven to ten year projection period. Economic cycles are not explicitly addressed by the valuation framework for models.[14]

Researchers around the world are stressing the need to perform market analysis to pay close attention to the local business cycle. Stephen Roulac suggested that changes that affect property cycles are not likely to be discovered by "traditional economic methods of extrapolating rents based on past patterns " As Roulac notes:

> Comprehending the new economy and how it drives the future real estate cycle involves the conceptual insights derived from assessing non-traditional information, tasks that largely involve creativity and qualitative interpretation, rather than traditional disciplined quantitative analysis.[15]

GIS is a very effective tool for keeping track of the business cycle, analyzing the likely effect of the cycle, and noting transformations that are altering the demand for real estate. Its capabilities have been demonstrated by example in this chapter.

In a recent Appraisal Institute publication, Robert E. McHolland, MAI, made the observation that many of us are basically using the same methodologies that were taught in the 1930s.[16] Sure, we are using computers instead of typewriters and we don't have to make trips to the recorder's office to find property sales, but we are often relying on the assumption that we can almost always find three or more comparable sales that will tell us what our subject property is worth.

Meanwhile, appraisers worldwide are being criticized for allowing excess development in many large cities in the late 1980s. Some seem to think all appraisers are capable of doing is taking

12. Rodney L Jeffries, "Discounted Cash Flow Valuations - Pitfalls, Standards & Solutions," a paper published by The University of Auckland, New Zealand, 1995
13. "Understanding the Property Cycle," a paper published by The Royal Institution of Chartered Surveyors, London, 1995.
14. Waldo L. Born and Stephen A. Phyrr, "Real Estate Valuation: The Effect of Market and Property Cycles," a paper published by the Lumpkin College of Business, Eastern Illinois University, Charleston, Illinois, and Davis Realty Partners, Inc. Austin, Texas, 1995.
15. Stephen E. Roulac, "Real Estate Market Cycles, Transformation Forces and Structural Change," in *The Journal of Real Estate Portfolio Management,* published by The American Real Estate Society, 1996.
16. Robert McHolland, MAI, "Do Appraisers Have a Future?" *The Appraisal Journal* (October 1994).

the net income from a property and dividing it by 10% or discounting it by 12% over time. They believe that appraisal is pretty much a waste of time and money.

The Roulac Group wrote a paper that briefly discussed due diligence for the twenty-first century.[17] It commented on some shortcomings in appraisal that will not be resolved until clients start demanding better work products and appraisers start to become better compensated. It also quoted an article describing a current perception of the appraisal process: "the appraisal process is becoming more and more a word processing activity that reproduces prior assignments and tries to get paid on the basis of a new product."[18] John W. Ross, executive vice president of the Appraisal Institute, recently noted that the accounting profession has encountered problems due to technological changes.[19] They have been successful in responding to these challenges by applying their expertise to other value-added services, such as consulting.

It seems like more is being expected of every profession, with business and professional practice being reinvented at a rapidly increasing rate. Almost any economic endeavor that still does things the way it did 50 years ago is scheduled for Schumpeter's creative destruction, which states that capitalism by its very nature constantly requires newer and better ways of doing things.[20]

Do appraisers have a future? We are not the only ones who are concerned. The Royal Institution of Chartered Surveyors has an annual Cutting Edge Conference at which researchers try to figure out new and better ways of analyzing real estate. It provided an excellent forum for my paper on the quantification of stigma,[21] and many people in Europe are very interested in this topic.

A recent book on brownfields[22] noted, "What is missing in this country are appraisers who can place a value on contaminated property 'as is'" and "only a handful of appraisers in the United States are willing and able to perform this key function in appraising brownfields." At lease one GIS program, EMAGIS, can be used to characterize the contamination in soils and provide an estimate of the resulting health risks.[23] These health risks are the causal factor for the financial risk of contaminated real estate. Financial risk needs to be quantified to value contaminated sites "as is."

SUMMARY AND CONCLUSION

Some of the most highly regarded real estate advisors have described what went wrong in the real estate markets during the late 1980s. They offered some suggestions to avoid problems in the future:

- Quantify the demand for real estate by determining the geographic area in which the demand for the property originates and
- Analyze and forecast employment by SIC categories for this area

Simple subjective judgments by investors are said to have been a problem. Such judgments have been the cause of wide divergence in opinions of value by appraisers for the same property, suggesting the need to quantify the demand for real estate in appraisal as well.

I have reviewed hundreds of appraisals of problem properties and discovered that many appraisals for the same property differed in their value conclusions by more than 30% for the same property at the same point in time. A local consultant noted that more than 100 shopping centers that had inadequate trade areas were reported to have been built in Phoenix during the late 1980s. I typically reviewed appraisal reports prepared after the bank that provided the construction lending failed. In one instance two appraisers referenced a ring study conducted to show the population around a shopping center that

17. Scott R. Muldavin and Stephen Roulac, "Due Diligence Enhancements for the 21st Century," a paper presented at the April 1998 meeting of the American Real Estate Society.
18. Allen Cyrot and Howard Zuckerman, "The Real Estate Industry's One-Dimension Paradigm Is About to Change," *Real Estate Development and Asset Management* (November 1997).
19. John W. Ross, "Responding to the Demands of the Marketplace," *Valuation Insights and Perspectives* (Second Quarter 1998).
20. Joseph A. Schumpeter, *Capitalism, Socialism and Democracy* (New York: Harper & Row): 83.
21. Weber, "Stigma - Unquantified Risk?"
22. Robert A. Simons, *Turning Brownfields into Greenbacks* (Washington, D.C.: The Urban Land Institute, 1998).
23. Bruce R. Weber, "Stigma: Quantifying Murphy's Law," *Urban Land* (June 1998): 12. This article summarizes a paper titled, "The Valuation of Contaminated Land," which won the first ULI manuscript award and is to be published later this year in the *Journal of Real Estate Research* published by the American Real Estate Society.

never attained more than a 50% rate of occupancy. Neither appraiser realized that it was a neighborhood center without a neighborhood. Both appraisers assumed (or made subjective judgments) that the center would readily attain stabilized occupancy of 95% within 18 months.

A number of the examples presented here illustrate problems associated with subjective judgments and demonstrate how they could have been resolved using GIS.

Example 1 shows that the use of GIS could prevent a property that does not have an adequate market area from being built. The problems associated with this property were not found by the use of conventional appraisal techniques. The example also illustrates invalid assumptions that are used in many appraisals—i.e., that the comparables all have similar trade areas and/or economic demand. GIS can be a solution to the problem because it can be used to quantify the demand for real estate that is generated by the surrounding population.

Examples 2 and 3 show that GIS can be used to expand the scope of services that an appraiser could provide. An appraiser using GIS could counsel a client as to the financial risk inherent in a contaminated site. GIS can be used to characterize the contamination and demonstrate visually if it is likely to result in significant health and financial risk. The appraiser would base this judgment on information from a soil scientist, rather than making subjective judgments that contaminated comparables can be used to reflect the financial risk of the subject.

Example 4 presents another problem that is difficult to solve without the use of GIS. Many of the appraisals that I have reviewed made subjective judgments that all mitigation land has equal utility. These appraisals did not discuss the fact that mitigation ratios can alter the desirability of what was thought to be generic mitigation land. GIS can also be used for relatively rapid spatial analysis to show the supply of and demand for land with this attribute. The example also illustrates the use of GIS for calculating land areas.

Example 5 refers to a situation in which appraisers made the subjective judgment that industrial land was the same over large geographic areas. GIS can be used to rapidly obtain an understanding of trends and segments in the market for industrial properties and other properties by the use of buffers. Geographic market segmentation can be difficult to determine with other methods.

Examples 6 and 7 show how GIS can be used to define the most relevant market area for a property and how sensitivity analysis can be conducted to check for demand within this market area once it has been defined. Actual market areas seldom end up being circles, as they are depicted. A review appraiser using GIS can provide much more meaningful appraisal reviews. These examples also show that the use of GIS can provide the creativity and qualitative interpretation that is said to be required to perceive changes in market conditions.

The business cycle forecast example illustrates how GIS can be used to find the market area from which most of the demand originates. A forecast of employment can then be made for this specific submarket. This method is superior to using subjective judgments that assume that the demand for the property will be the same as the demand reported for other properties in a national survey.

Finally, GIS can help appraisers respond to the criticism they have received in the past. Pension fund advisors are using GIS, not subjective judgments, to quantify the demand for real estate. If appraisers emulate the market by using GIS as well, they might be able to convince their clients that they do not have to learn how to use this tool themselves to obtain a thorough analysis of economic demand. At the very least, GIS is a good supplement to traditional quantitative analysis. Hopefully its use will enable appraisers to provide additional value-added services to their clients.

chapter 6

The Appraiser and GIS in a Litigation Context

by Bill Mundy, MAI, PhD, CRE

This chapter addresses the role of an appraiser as an expert witness and how an appraiser can deliver better testimony through the use of GIS. An actual case in which GIS was used is discussed.[1]

THE CHOICE OF AN EXPERT

Attorneys often use three criteria in selecting an expert witness:

1. The individual's experience and qualifications
2. Age, which is one of the few situations in which baldness, gray hair, or wrinkles really do help
3. Teaching ability and communication skills, which indicate the expert's ability to communicate effectively in court

All of these attributes can be important in determining the "prestige factor" associated with the expert. The importance of this factor was demonstrated in a recent decision where the judge assumed the responsibility of a "gatekeeper," making sure that the experts called really were experts so that court time would not be wasted on irrelevant matters or witnesses.[2] With regard to teaching ability and communication skills, a good-quality GIS presentation can help significantly in communicating essential information to the court. Visual materials make the expert's testimony easier to comprehend and more interesting.

A GIS presentation can introduce realism into the courtroom. It can make a difficult subject come alive so the court both understands and enjoys the testimony. The expert needs to simplify background data as much as possible, yet convey its meaning clearly. This can be done with good-quality illustrations.

THE APPRAISER AS AN EXPERT

The appraiser can use his or her GIS expertise to help the attorney and property owner from the inception to the completion of the case.

1. The case in question was settled shortly before trial, partially because of the effective use of GIS.
2. See "Admissibility of Expert Testimony After Daubert: The 'Prestige' Factor." 43 *Emory Law Journal* 867 (1994).

Inception

The attorney must have a sound understanding of the nature of the property and its neighborhood. The appraiser may provide key information on value relationships and other influences related to litigation matters. A statistical analysis of important value information and its graphic presentation in charts and on maps can assist both the valuer and the attorney in understanding the nature of the problem at hand.

For example, consider a class action lawsuit that involves airborne contaminants. The appraiser is working with the law firm to help prepare the class certification, which involves identification of the geographic area affected. The appraiser can use GIS to plot the occurrence of air quality complaints by households and supplement these data with information on wind patterns. Then isolines can be developed to indicate the extent of the problem and how the intensity of the problem varies with distance from the source.

Another aspect of a class action lawsuit is the homogeneity or diversity of the properties considered in the suit. Generally, the more diverse the properties in the class area, the more difficult it is to achieve class certification. By incorporating assessor information on land use within the geographical area, the appraiser can determine how homogeneous the land use types are. Some geographic pockets may be eliminated, e.g., those with significant tracts of nonresidential property. When plotted, property diversity is easy to identify.

Data Gathering and Analysis

Sometimes the gathering and analysis of data for litigation can be relatively simple. For example, in valuing a particular property—say, a strip commercial center—the appraiser concentrates on data assemblages and analysis for that property only. In other situations, the gathering and analysis of data can be very complex—e.g., in a class action lawsuit where 5,000 properties are involved in the litigation. For a typical appraisal, a simple, geographic-based system can be used to provide good visual aids and a more presentable report. Figure 6.1 is an example of a simple map produced using DeLorme mapping software.

For a major lawsuit, on the other hand, an entire database may need to be established to help the appraiser determine accurately and efficiently the existence and extent of any value-influencing factors. Determining whether value impacts are present can involve numerous types of statistical analyses, including paired sales analysis, multiple regression analysis, and surveys of neighborhood groups.

Example

The following discussion relates to a recent major lawsuit involving 6,344 separate residential properties located where dangerous emissions from a smokestack caused contamination problems. A comprehensive database and mapping system of the entire class action neighborhood was developed. Each property was identified by address, which provided its longitude and latitude. Assessor and multiple listing service data were imported into each property's record file. In this case, the hypothesis was that proximity to the emission source (i.e., the smokestack) would have an influence on property values, assuming that buyers were knowledgeable and adequately informed about the nature of the contamination problem. The neighborhood map, boundaries of the class area, and land use data are shown in Figure 6.2.

Through research, the appraiser found that damage caused by contamination from the smokestack varied as one moved away from the emission source. An Atlas GIS system was used to calculate the amount of damage assigned to each residence. In a study area within approximately one-half mile of the stack, the typical decrease in the value of a home was $24,950. Beyond that area, but still within the class action area, the damage dropped to $16,700. The damage continued beyond the class area, but because the people who lived beyond the class area were not included as a part of the litigation, they were not entitled to any compensation as a part of this suit.[3]

3. This is a good example of how preliminary analysis (before class certification) can significantly assist the attorney/client who has retained the appraiser. In this case, the class area should have been larger.

Figure 6.1
Simple Software-Produced Map

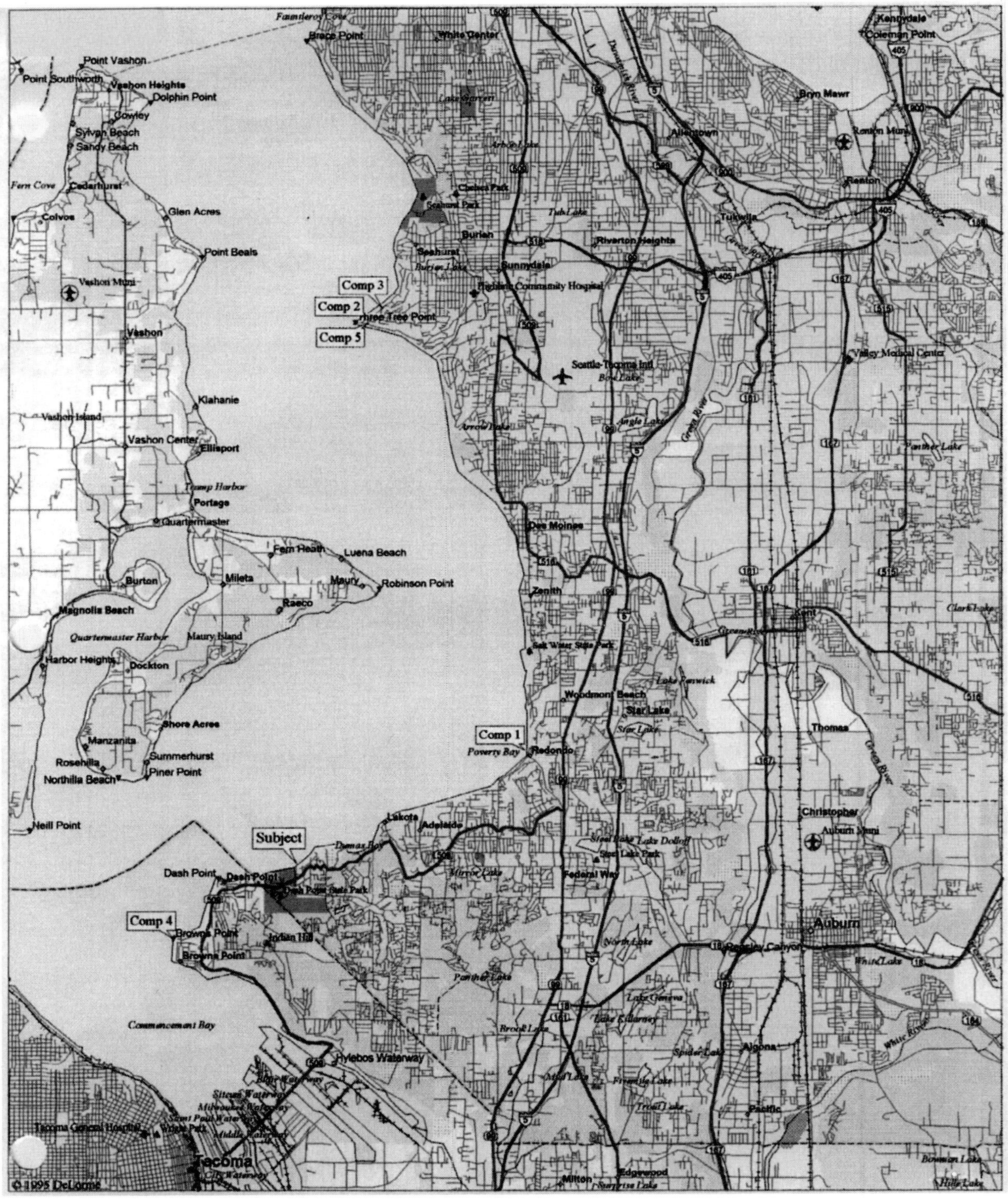

A color copy of this figure can be found on page 205.

Figure 6.2
Neighborhood Map

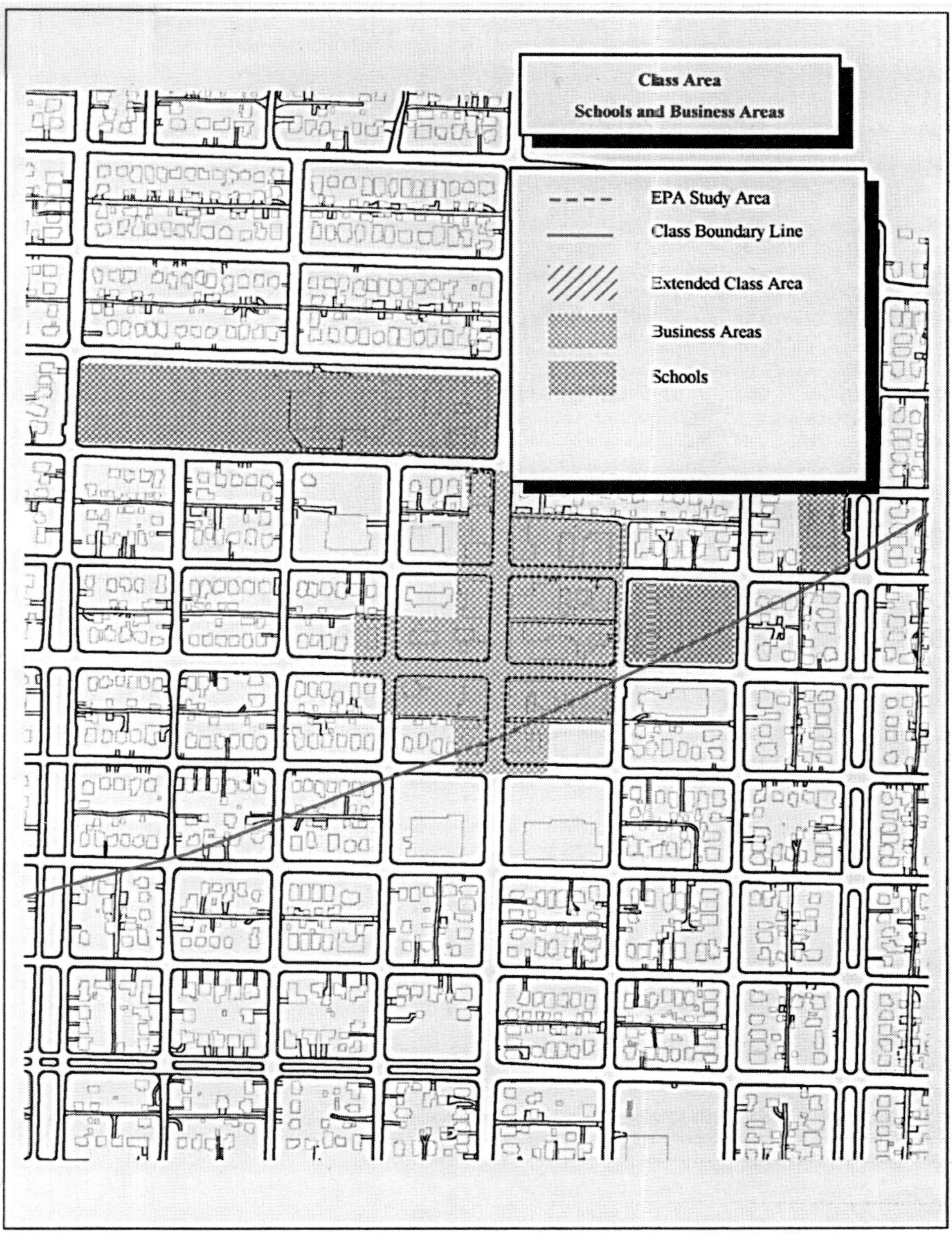

A color copy of this figure can be found on page 206.

The value loss for a typical $80,000 home is expressed by the following formula:

$$Loss_{80} = b + (m)D_{stack}$$

where: Loss = Loss in the value of an $80,000 home

b = a constant

m = slope

D_{stack} = distance from smokestack

Next, a value decline function was calculated. This consisted of two variables. The first is the constant, or intercept, which is identified as *b,* while the second is the slope of the value function line, identified as *m.*

The slope of the function (m) can be calculated with the following formula:

$$m = (Y_2 - Y_1) / (X_2 - X_1)$$

where: Y_1 = damage in study area = $24,950

Y_2 = damage in area beyond study area = $16,700

X_1 = iso one radius = 0.685

X_2 = iso two radius = 1.649

Solving for m yields a value of -$8,558. Therefore, the slope of the function is negative, indicating that the amount of damage decreases by $8,558 per mile as one moves away from the stack.

Next, we solve for the intercept b using the following formula:

$$b = Y_1 - mX_1$$

where: Y_1 = damage in the study area = $24,950

m = the slope = -$8,558

X_1 = iso one = 0.685

Solving for b, the value of damage at the stack is $30,812. This is the intercept. The value decline function is plotted in Figure 6.3. The data for the function is shown in Table 6.1. Table 6.2 shows how damage was calculated.

Table 6.1

Loss of Value with Distance

Distance (*D*)	**Loss for $80,000 Home = *b* + *m*(*D*)**
0.00	$30,812
0.25	$28,673
0.50	$26,533
0.75	$24,394
1.00	$22,254
1.25	$20,115
1.50	$17,975
1.75	$15,836
2.00	$13,696
2.25	$11,557
2.50	$9,417
2.75	$7,278
3.00	$5,138

Figure 6.3
Value Decline Function

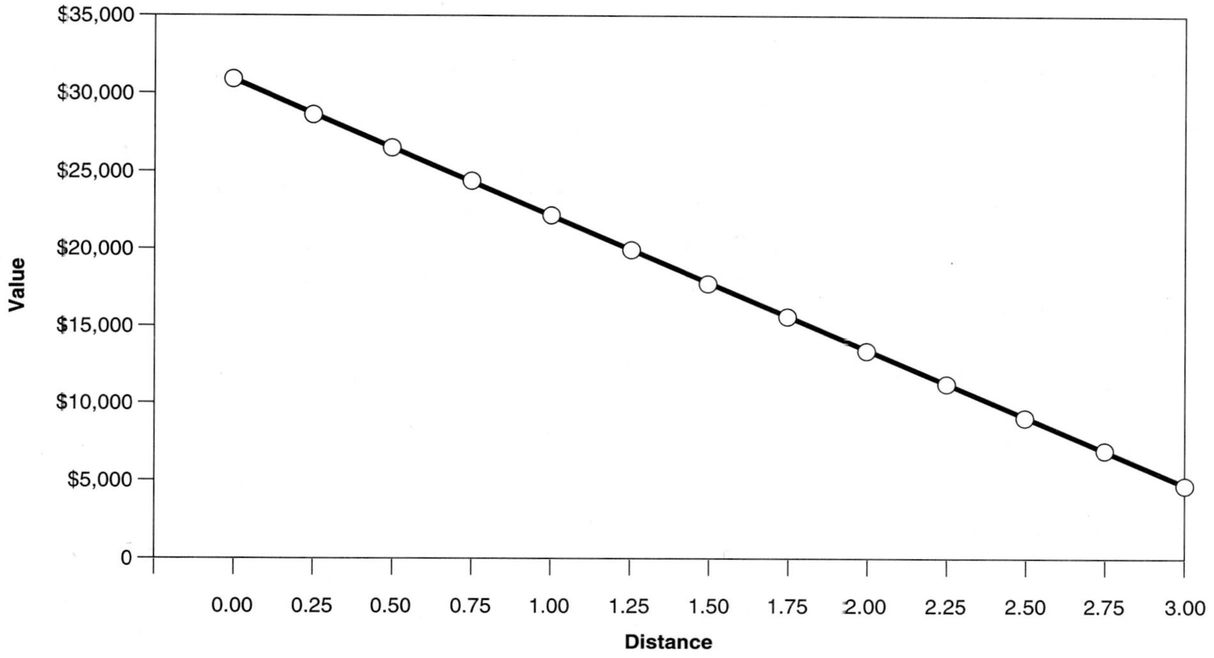

Table 6.2
Loss of Home Value

Distance *(D)*	Assessed Value	Market Value *(MV)*[5]	Damage
0.00	$80,700	$96,840	$37,298
0.25	$60,100	$72,120	$25,849
0.50	$172,000	$206,400	$68,456
0.75	$180,300	$216,360	$65,973
1.00	$115,400	$138,480	$38,522
1.25	$73,100	$87,720	$22,056
1.50	$104,300	$125,160	$28,122
1.75	$119,100	$142,920	$28,290
2.00	$94,700	$113,640	$19,455
2.25	$120,000	$144,000	$20,802
2.50	$125,000	$150,000	$17,657
2.75	$130,000	$156,000	$14,191
3.00	$135,000	$162,000	$10,404

The second column in Table 6.2 lists the assessed value of a sampling of homes. The analyst checked the accuracy of the assessor's information against actual sales. The selling price of each home was compared to its assessed value. On average, market values were 120% of assessed values, so assessed values were adjusted upward to reflect market value. The last column in Table 6.2 shows the damage based on the value decline function for a cross-section of the homes. The market values *(mv)* were adjusted from the $80,000 standard. The final equation becomes

$$\text{Loss}_{mv} = (mv/80{,}000)(b + mD)$$

All of the 6,334 homes in the class area were different distances from the smokestack. The distances are given in Table 6.1 and 6.2. An algorithm was written into the GIS software to calculate the distance between the smokestack and each home. This was accomplished using the address of the home converted into a longitude and latitude. The longitude and latitude of the smokestack was also determined. The Atlas GIS software calculated the distance between the two points, each house and the stack. The distance *(D)* was inserted in the value decline function and the damage for all the homes was calculated.

This calculation was performed in a matter of seconds. Manually it would have taken months because the distance between the smokestack and each house would have had to be calculated and entered into a spreadsheet to calculate individual damages.

SUMMARY

The example presented has demonstrated how GIS can be used

- As an effective tool in litigation valuation work to improve both accuracy and efficiency
- To create graphics and facilitate the presentation of data in a trial
- In database management to improve spreadsheet analysis and to perform necessary calculations efficiently

At Mundy & Associates we have found GIS to be extremely effective in improving the efficiency of our valuation analyses. We are currently using GIS in

- A class action project that involves 5,000 urban fringe properties in one state
- A major class action encompassing some 65,000 urban properties (homes, commercial buildings, industrial, vacant land) in another state
- A preservation and conservation litigation matter in a third state, which involves some 500 properties averaging approximately 600 acres in size that have unique features relating to wildlife habitat, archaeology, and paleontology

Without the efficiency afforded by GIS, we would not be able to handle projects of this size with our small, 11-person staff. Our current GIS system consists of the following hardware and software:

Hardware: Pentium 2,400 MHz processor-based PC
128MB RAM
19" SVGA monitor
32X CD-ROM
Open GL compliant graphics card
HP Deskjet 870cxi printer
Summagraphics Summagrid IV digitizer (36" x 48")
Toshiba Satellite Pro Laptop (P133, 32MB)
Toshiba Libretto sub-notebook
Magellan GPS receiver

Software: ESRI (Redlands, CA) ArcView ver 3.0a
ESRI StreetMap
Caliper (Newton, MA) Maptitude ver 4.0
DeLorme StreetAtlas 5.0 with Tripmate GPS

chapter 7

Software, Data, and Management Considerations in Establishing a GIS/Mapping System

by L. Ried Schott, MAI, SRA

Most business people taking the plunge into geographic information systems (GIS) are daunted by the technical vocabulary or steep learning curve involved. Unfortunately, their fears are justified. As this chapter will attest, a great many factors must be considered in the implementation of a GIS. What you absolutely need to know depends on what you want to accomplish and what you already know about computers, real estate, and GIS. There is no single, simple solution that will suit everyone's needs, but some applications are much less complicated than others.

Computerized mapping is much simpler than GIS and can accomplish many basic GIS functions. After reading this chapter, you can decide which option, GIS or mapping, best suits your needs. Another alternative for business people who do not want to get too technical is to outsource some of their more difficult GIS tasks to consultants or value-added resellers (VARs). For extremely complex projects, it may be best for a system integrator (SI) to handle almost everything.

This chapter provides an overview of basic considerations for establishing a GIS or mapping system. It begins with an examination of the conceptual differences and similarities between mapping and GIS and the critical factors necessary for GIS and mapping to be successful—especially software and data. Management issues are examined in detail. These include financial considerations, vendor selection and support, staffing and assistance, staff education and training, and implementation alternatives. Indirect considerations are often overlooked, so preimplementation studies and organization are essential.

SIMILARITIES AND DIFFERENCES BETWEEN MAPPING AND GIS

Most of the literature to date has failed to distinguish between computerized mapping and GIS. Index services sometimes classify GIS under mapping because the two can present similar images and printouts. They both deal with geography and can find locations, even by street address. Nevertheless, the two systems have distinct differences.

GIS is mapping but with more intelligence. Mapping software shows you what it has and can give you limited information about cities, states, or places. GIS shows you what you want using information that you decide it should have. The difference is similar to having a meal delivered from a take-out menu versus making a meal from scratch. You can make yourself just about anything, but menus limit your options to those selected by another person.

GIS can handle multiple levels of information simultaneously. If you want to see collectively which cities or neighborhoods have, say, a specific range of per capita incomes, a maximum proportion in an age group, and a certain minimum average housing value, GIS can do this. Traditionally, mapping cannot. Mapping has approximate locations and distances, while GIS has precise geographic coordinates and exact spatial distances. The ability to interrelate multiple layers of information spatially is an extremely powerful capability that geographic information systems has and computerized mapping does not.

Another feature that distinguishes GIS programs is their fundamental design, which is either vector- or raster-based. *Vector* refers to linear connections between x,y coordinates and a vector-based program produces maps composed of connected points, lines, and areas (polygons). Vector-based GIS uses a topological data structure in which elements of a map are defined in such way as to know which line segments are connected to each other and which polygon is adjacent to the side of each line. In contrast, raster maps have a grid or matrix value structure that allows for the organization, classification, and identification of different types of data.

Computer speed can be increased when using raster maps, but the ability to zoom in and out is more limited. Vector-based maps usually have a slower redraw time but have better zoom capabilities. The distinctions between raster- and vector-based maps are becoming blurred as programs accommodating both structures are developed and redesigned. With the combination of the two structures the user can, for example, draw zip codes and streets over a raster map (such as an aerial photo), thereby improving visualization and increasing the system's capabilities. More specifically, users may ask: What can I actually accomplish with one system or the other?

With computerized mapping, a real estate broker or appraiser can easily prepare location maps for a specific property and comparable sales or listings. Property addresses can be imported from a list in a word processor, spreadsheet, or MLS inquiry and automatically located on a map. Moreover, some mapping programs can provide directions and show the best travel routes to a property or even several properties. The capacity to identify a property and obtain directions for travel between two locations is available free on the World Wide Web (the Web). Such maps can be used by most real estate professionals, whether they are appraisers, brokers, mortgage lenders, or others. The maps can be given to clients as a separate item or within a report. If the appraiser constantly needs to prepare maps or inspect properties, mapping products can save time.

With GIS you can ostensibly do the same mapping functions, but GIS allows you to include additional information that can be used to screen the data or help you better understand given areas. For example, school districts and flood zones can be added on to a map, which may help explain why some properties sell for more than others. While mapping provides a solution to a specific problem, such as locating properties on a map, GIS provides access to multiple types of information simultaneously. This may allow the analyst to discern trends affecting absorption, rents, values, population growth, etc. GIS facilitates a more scientific examination of any relationship that can be mapped. This ability for the real estate industry and practitioners to analyze and understand real estate in a new way may be the single most significant phenomena to ever impact real estate analysis.

The introduction of GIS into real estate has been slow, largely due to the way that GIS products have been developed. GIS and mapping products have emerged from different types of companies with different perspectives. The primary focus of GIS companies has been on natural resources, marketing, transportation, government, and general business fields. In contrast, mapping companies focus on education, navigation, consumer assistance, and business travel. Interest in real estate from either group has been relatively recent and has a long way to go. Most of the work in developing a GIS real estate system has been left to users or their paid consultants. GIS products customized for real estate by resellers are being introduced relatively slowly, and those currently available may not meet your specific needs.

The demand for mapping and GIS capabilities is primarily coming from two different directions. High-end GIS applications are needed to solve multiple problems such as facility management in large companies or tax assessment and zoning in local governments. Smaller users want solutions to specific mapping needs. The middle market for GIS applications is also developing, in particular with VARs and consultants. Garnishing the advantages of GIS involves a much greater commitment of time, resources, and money than implementing a mapping system.

CRITICAL SUCCESS FACTORS

To implement either a mapping system or a GIS successfully, one must consider several factors:

- Solution to a problem, provide added benefits
- Easy to learn and productive to use
- Ability to procure relevant data and ease of access
- Expandability
- Support
- Linkages to MIS and applications
- Fast payback and limited risk

To solve problems and provide added benefits, it is vital to identify tangible and realistic goals. Some GIS applications have been presented earlier in this book. A large number of other problem-solving applications for GIS and mapping have not been mentioned and may not even have been thought of yet. Knowing more about the capabilities of these programs will help you to determine which you need and to define your goals. The benefits of GIS beyond mapping can be numerous—e.g., increased business and productivity, new services, reduced costs, improved analytical capabilities, loss avoidance, and a more professional image.

The ease of learning and the productivity of computerized mapping and GIS software are critical factors. A well-documented manual, a good tutorial, and a helpful training session can shorten the learning curve of any software considerably. The software must have an intuitive feel to make it easy to use. The transition of GIS software from a DOS-based system to Microsoft's Windows with more standard menus and protocols facilitates learning. Customization features, available through macros or a scripting language, are also essential if specific needs are to be satisfied. Ease of learning and productivity apply to hardware and data sources as well. Some hardware operates more easily and quickly than others. Data sources that come with better documentation and have added filtering features enhance productivity.

The procurement of relevant data serves as the foundation for any GIS system. Data may pertain to boundaries, demographics, economics, real estate, or images (e.g., aerial photos, diagrams, plans). The data procured depend on the user's needs. For GIS, in-house or primary data are normally the basis for a system. Other data files for streets or county boundaries, for example, are usually necessary, as are demographic, economic, or real estate data. It can be helpful if relevant data files come bundled with the GIS program; this is usually mandatory for a mapping program. The GIS program selected should be able to accommodate files from a large number of possible suppliers. The importance of data, particularly for GIS, relates to the software's ability to handle different types and quantities of data.

Expandability depends on hardware and, to a lesser extent, on software. Expansion allows you to keep up with the increasing demands software makes on your system. When budgets are tight, choosing a system that can be expanded will often pay off in the future. You may not be able to afford certain items now, but later when your financial situation may be less constrained, you can add on those items. For example, it may be wise to start out with a computer that has a powerful microprocessor and more drive bays and expansion slots rather than an underpowered computer with a much larger hard disk. Another hard disk or storage device can easily be added at a later date, for relatively little cost. Some hardware components are much easier to upgrade than others, as will be explained later in this chapter. If many hardware devices will be required, a system with a SCSI interface may be desired.

Expandability is also important for add-on hardware or peripherals. The ability to increase printer memory, for example, may allow the printer to produce more complicated maps. If a larger monitor will be needed later, you may want to invest in a better video card initially to avoid purchasing two items—one now and an upgraded one in the future. Likewise, software expandability is essential. If you are certain that you only require a program for a specific use, then mapping software alone may be fine. However, it is advisable to consider programs that have various add-on products or an upgrade path with advanced products that will accommodate your growing needs.

Support becomes more important as the complexity of software, hardware, or data continue to increase. The need for support is usually inversely related to the expertise of the user; the less the user knows about a particular system component, the more vital support becomes. Support is usually obtained from suppliers and/or vendors, often over the phone. Technical support can also be provided through bulletin boards, fax services, the Internet, e-mail, and other means. In the past support was often free, at least for a limited period, but the phone call required a long distance charge. Increasingly, however, support is provided for a fee. Related matters to be investigated include warranties, policies and agreements, staffing and assistance, corporate support in an organizational environment, education, and training. Some vendors provide training sessions while others do not, and training on certain products may be difficult to obtain. Given the wide scope of factors affected, support is not an issue to be ignored.

Linkages or connectivity and data compatibility are other significant issues, especially if the GIS is to be used by multiple groups or offices. Even if there is only one GIS user, productivity can be hampered if the data needed cannot be directly accessed and used in their existing format or easily imported. Constant exchanges of data and the need for conversions consume valuable time. Connectivity or linkage is a function of direct components, such as hardware and software, but it is also affected by indirect system component considerations, such as vendors, who may or may not support network capabilities, education, staffing, and assistance.

The payback/risk of GIS or mapping must be assessed. The simpler the system, the less risk is involved and the faster the payback. Buyers of more complex systems may pay a greater price, which may or may not be surpassed by the ultimate benefits, but a longer payback period is usually involved. To minimize the risks, different approaches can be used. One might begin with an inexpensive mapping system. Another option is to start with limited GIS. Limited GIS could begin on a stand-alone system and later spread to a work group and other departments, following a set implementation plan. With a more complex system, timing is likely to depend on how quickly data and assistance from other departments can be coordinated and various organizational issues can be addressed. Staffing, assistance, and education requirements must be added to your list of concerns to be considered when implementing a GIS.

Geographic information systems are an emerging technology. Some users believe the benefits of GIS are enormous. Although certain capabilities remain intangible wishes, others have become a reality. Since advances in this field are occurring rapidly, systems that were state-of-the-art a year or two ago may be obsolete today. In terms of the following discussion, readers should keep in mind that innovations become dated very quickly.

SOFTWARE

Background

GIS and mapping software programs vary considerably and usually contain limited data. In particular, few programs contain real estate-specific data. Since software and data are the heart of a GIS or mapping system, it is vital that numerous software and data issues be considered when developing such systems.

The advent of commercial GIS software can be linked to the creation of the Environmental Systems Research Institute, Inc. (ESRI) in the late 1960s. Demand for sophisticated mapping and the informational needs of governmental, scientific, and academic interests propelled this firm's flagship product, ARC/INFO, to the forefront of GIS. This and similar upper-end GIS products, from Intergraph, Tydac, and others, were initially used on mainframes and mini-computers and most recently on high-end file servers.

Now, a quarter century after ESRI's founding, mid-level and lower-end mapping programs for personal computers are flooding onto the market, bringing along desktop GIS products that offer general and sophisticated capabilities. In the mid-1990s the market for GIS programs became ripe and now offers programs to better manage and understand information for better decision making. With advances in computer technology, PCs can better satisfy the extensive computer power and memory requirements of mapping programs at increasingly affordable prices.

Spreadsheet software from Lotus helped make mapping accessible to personal and business computer users in 1993 with Lotus Corporation's integration of a mapping product in 1-2-3, Version 5. Since most of the data used is geographically oriented, it was only logical that mapping software was eventually introduced into spreadsheet software. A year later, Microsoft added a similar component into its spreadsheet Excel 4. Neither Lotus nor Microsoft had mapping expertise, so they needed to affiliate with other companies. Lotus associated with SMI/Atlas GIS and Microsoft teamed up with MapInfo.

At about the same time that Microsoft affiliated with MapInfo, Microsoft quietly acquired a mapping company by the name of Automap, Inc., located in Bellevue, Washington. Although Microsoft began marketing its newly acquired Automap products as part of its "Home" products in 1995, a more robust Automap Streets Plus bundled with a yellow pages directory appeared in 1996. Shortly thereafter, it became part of Office 97's Small Business Solutions Suite, which includes Word, Excel, Publisher, Outlook, and more. This marks the first time a mapping program was featured for business use by a major software company along with its flagship products. In 1997 the product name changed to Expedia Streets and Microsoft introduced mapping capabilities on the Web known as ExpediaMaps. The Expedia name reflects an affiliation with Microsoft's Travel Agency. Mapping's inclusion in mainstream software and its use on the Web for both personal and business use signifies its increasing importance.

No longer the exclusive domain of large computer systems, GIS and mapping have been brought down to the desktop, the laptop, and even the palm-sized computer.

GIS Categories

GIS and mapping can be organized into five categories. The major categories under GIS are

- High-end
- Mid-level
- Entry-level

The categories under mapping are

- Business and personal
- Home

These categories are depicted in Figure 7.1 and described in the following section.

Due to greater functionality, GIS products have much broader applications than mapping products. All three GIS categories are more powerful than the two mapping categories.

High-end GIS

High-end GIS is the most sophisticated. It is full-featured and has the most proven set of programs. Originally developed to run on mainframes or minicomputers, these programs are now well served by high-level file servers and workstations. They ordinarily run on a Unix or upper-end operating system

Figure 7.1
Categories of GIS and Mapping

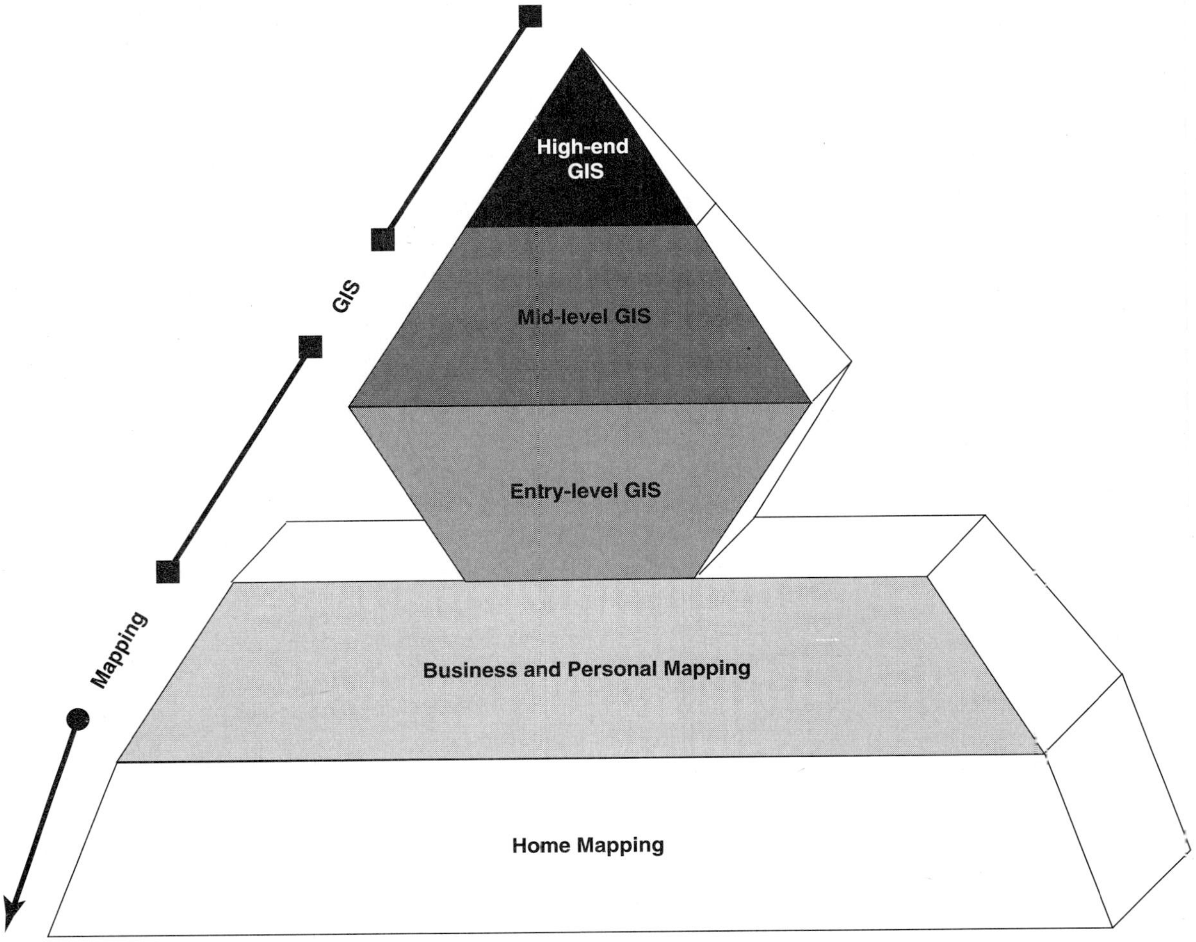

Source: Schott Consulting Services

and are able to serve a large number of users simultaneously. Although personal computers are limited in their ability to tackle high-end GIS, they can often be linked with more powerful computers through local or wide-area networks accessing certain features of such programs. A GIS product with fewer features (i.e., mid-level or entry-level GIS) can be used on a PC to access and exchange information from a high-end GIS as well. Thus, what was once available only to workstations and large computer systems has become, thanks to improvements in technology, accessible to personal computers.

In addition to the ability to run on servers and strong networking capabilities, high-end GIS features are distinguished by

- Three-dimensional capabilities, which can be used for topographic features, for example
- A powerful programming or scripting language
- Sophisticated analytics
- The ability to interact with large database systems

These capabilities are derived in part from the programs' initial use in scientific, academic, and government fields.

High-end GIS has been applied to the government, academics, natural resources, and business. It best serves applications that require large quantities of data, maps, plans, and photos; accommodate a large number of users; or integrate data with other departments.

Historically, high-level GIS has not been considered user-friendly. It may require months or even years of training. Some universities offer degrees or concentrated programs of study in GIS, principally to achieve understanding and proficiency in the operation of high-level GIS systems. GIS education and learning is not confined to academic institutions. The major software vendors provide classes and there are numerous books available on the subject. One Internet source for such books is RealtyBooks[SM] (http://www.realtybooks.com).

Despite their broad capabilities, buyers of high-end GIS products may need to acquire various add-on modules from vendors to obtain a complete set of features. Often these programs are stripped down for better price appeal and to attract users who require fewer highly sophisticated features. Still, these products are fairly expensive, costing at least $2,000 and up to more than $15,000 with various add-on products. Moreover, as the number of users increases, the cost also rises.

Several companies provide high-end GIS programs, including Environmental Systems Research Institute (ESRI) and Intergraph Corporation.

Mid-level GIS

Mid-level GIS products have taken advantage of GUI and have become easier to use and learn. They have become increasingly popular due to this improvement and to the advent of less-expensive, more powerful computers, a reduction in the price of the programs, and data bundling.

Mid-level GIS programs operate on high-end, stand-alone desktop computers; they can also operate on networks, albeit with varying success. These programs are constantly improving, and many now provide most of the capabilities included in much more costly high-end programs.

In the past mid-level programs were typically not bundled with data. This changed in 1994 when First Street, the first product to bundle extensive boundary, street, and census data and a GIS program (ArcView), was introduced by Wessex. About a year later, Calipers became the first GIS company to bundle its GIS product—Maptitude—internally with similar data files.

In the future more mid-level GIS programs will include boundary and/or demographic bundles, which should help increase sales and application development. Also, some products in this category may be retargeted or discounted.

Mid-level GIS can be considerably easier to use than a high-level GIS. Rather than a program of study at a university, training for mid-level GIS can often be accomplished in a course or two. It is often taught in the geography department of a college or at a certificate program. Software companies also offer courses in mid-level GIS, and there are books on the more popular programs.

In the early 1990s mid-level products, including extensive data and boundary files, could easily cost $10,000 to $50,000. Several years later, this cost has dropped dramatically and is typically between $300 and $1,500 with some bundling of data.

Many companies provide mid-level GIS programs as well as other programs that are less or more sophisticated. Such products include Tactician by Tactics International, Maptitude by Calipers, ArcView and Atlas GIS by ESRI, and MapInfo Professional by MapInfo. The GIS companies mentioned and the mapping companies discussed below, along with their products and price categories, are shown in Table 7.1. An updated list is accessible via the Internet at InRealty[SM] (http://www.inrealty.com/geolists.html).

Entry-level GIS

As indicated by its name, entry-level GIS possesses fewer features than other programs and may be considered GIS-lite. Even so, some of these programs come bundled with a great deal of examples and/or data and have many capabilities.

Although some scaled-down mid-level GIS versions were introduced in the mid-1990s without much success, newer, more innovative entry-level GIS products are evolving. Larger GIS companies have products that provide a bridge from mapping to GIS and some use "mapping" in the product name instead of GIS to avoid scaring potential customers. The distinction between entry-level GIS programs and upper-level business mapping programs is not always clear.

Table 7.1
Leading GIS Software Vendors

Price Category	Company	Product
Over $5,000 (GIS)	Intergraph	MGE
	ESRI	Arc/Info
	ESRI	MapObjects *et al**
	Intergraph	GeoMedia*
	Autodesk	MapGuide*
	MapInfo	Proserver, *et al**
Less than $5,000 (GIS)	MapInfo Corporation	MapInfo
	ESRI	ArcView
	ESRI	Atlas GIS
	ESRI	MapObjects
	Autodesk	AutoCAD
	Caliper Corporation	Maptitude
	Tactics International	Tactician
	Equifax, *etc.*	Infomark, *etc.*
	Scan/USA, *etc.*	Scan/USA, *etc.*
Under $300 (Mapping)	DeLorme	Street Atlas USA
	ESRI	BusinessMap
Under $100 (Mapping)	MapLinx Corp.	MapLinx
	RandMcNally	StreetFinder
	Microsoft	Expedia Streets*
	Thomas Bros. Maps	The Thomas Guide
	Vicinity, mapQuest, *etc.*	MapBlast, *etc.*

*** Internet capable. This capability is increasing among most programs.**

Source: Castle Consulting & Schott Consulting Services.
Note: The product names and price categories noted above are subject to change. An effort by the author to provide current company and product information can be found on the Internet at InRealty[SM] (http: www.inrealty.com/geolists.html).

Entry-level products share certain characteristics with mapping programs. First, they are not usually the ultimate end-product offered by the vendor. Second, they often include some data bundles that provide a lot for the money, but still require add-on files. These additional files may cost much more than the original program. Third, because they lack many of the more complicated features of a mid-level GIS, these programs can be easier to learn and use.

Entry-level GIS programs usually cost one-half the price of mid-level GIS programs or less. Prices begin at a few hundred dollars for programs that consist of upgraded mapping programs with more data files and add-ons. Some programs in this category are basically scaled-down mid-level GIS programs to which increased features and data can be added later. Ultimately, the combined prices of add-ons, additional data, and the entry-level product may approximate the cost of a mid-level GIS program. Companies offering entry-level products include MapInfo and Scan/USA.

Table 7.2 lists GIS and mapping features.

Mapping Categories

Two categories of mapping are discussed here: business and personal mapping and home mapping. The personal aspect of mapping can also be separate or may be included in the home category. However, at this juncture it appears that business and personal mapping are closely aligned and are therefore best considered in tandem.

There are many inexpensive and less well-known mapping programs. The following discussion is not a comprehensive list of individual mapping programs because the names of products

Table 7.2

GIS/Mapping Features

- **Basic operations**
  Zooming and panning
  Multiple layers
  Thematic capabilities
  Mapping symbols
  General querying capabilities

- **Map modification and creation**
  Creation capabilities
  Tools available
  Input from digitizers and scanners

- **Spatial queries**
  Pointed and weighted aggregation
  Built-in value

- **Database management and formulas**
  Storage space
  Sorting capabilities
  Merging capabilities
  Arithmetic functions

- **Graphics**

- **Transfers and display**
  Importing and exporting
  SQL transfers
  Embedding (DDE/OLE)
  Multiple windows

- **Reporting and printing**
  WYSIWYG
  Templates
  Printer compatibility and capabilities

- **Networkability**

- **Speed**
  Opening and closing files, maps, etc.
  Finding addresses
  Redrawing maps (with streets)
  Zooming and panning
  Geo-coding
  Combining multiple layers of maps
  Performing complex queries
  Performing database functions
  Printing maps and reports

- **Expandability**
  Peripheral support
  Cross-platforms
  Customization

- **Documentation/tutorial**
  Manual
  On-line services

- **Ease of use**
  Program complexity
  Extent of documentation
  Tutorials

- **Customer support**
  Cost
  Hours of operation
  Accessibility
  Expertise of support staff
  Update and exchange policies

- **Maps and data provided**
  General bundles
  Customized packages
  One source shopping

and the companies that make them change frequently. Instead, an overview of current mapping categories is presented.

Business and Personal Mapping

Demographic programs Various demographic mapping programs are suitable to multiple industries and services. These programs are usually bundled with national data but can also be regionally oriented. They provide both historical and current forecasts as well as mapping capabilities.

These products print standardized demographic reports, which are frequently used by retailers, marketing people, and real estate professionals. Such a report might include demographics for one-, three-, and five-mile concentric rings around a specific subject property. Demographic information on age, sex, population, income, education, race, and other characteristics is broken down for each of the three areas. A map displaying the concentric areas can accompany the report. Additional thematic maps of the demographic variables cited are also available.

Relatively inexpensive programs for desktop computers (costing $100 to $1,000) include AnySite from Integrated Technologies, Business Map from ESRI, and Scan/USA and MapLinx from companies of those same names. Major socioeconomic data vendors such as Claritas also bundle mapping programs with their data, but these products can be very expensive ($15,000+). They derive most of their value-added from the data rather than the software and can be transcended by importing the vendor's data into a standard GIS.

Mapping programs for locating properties (geocoding) Most real estate services require maps to indicate the location of properties. To create a map, many real estate services copy a paper map, cut it up, paste it together, identify properties using peel-off, alphanumeric stencils, and enlarge or reduce the map on a copy machine. This procedure is tedious and time-consuming, but it has been used for decades.

Beginning in the mid-1990s, very affordable mapping programs became available. These programs offer specific, easy-to-use applications that can be used to create maps for different real estate purposes at relatively low cost. Some of the programs allow several properties to be automatically located. They can be annotated with countless symbols or numbers. Such programs are useful to residential and commercial brokers, appraisers, leasing agents/tenant representatives, mortgage lenders, management agents, and many other real estate professionals.

Microsoft introduced a revised "Internet-ready" version of AutoMap Streets Plus in 1996. This added a new dimension to mapping by providing information about a property via the Internet by simply clicking on an icon or symbol. This feature has numerous applications. For example, by clicking a "property for sale" symbol, descriptive information about a house or commercial building is readily obtained. Additional material can also be obtained, including pictures, video, floor plans, and e-mail links to the broker or owner. The user can obtain information on rental buildings (rates and amenities), buildings owned or managed (sizes and occupancies), employers (type of business, number of employees, and available jobs), and so forth.

If hundreds or even thousands of properties must be located on maps, mapping programs are often inadequate. GIS programs, however, excel in their ability to handle a great deal of data and complexity. They often use a database file containing property addresses as well as other relevant data and then geocode the addresses by assigning a geographic placement (latitude and longitude) to each of the properties. Some programs require an additional add-on product to perform geocoding. The ability to locate large numbers of properties is useful for financial institutions, corporations, government agencies, brokerage firms, management and leasing companies, and others.

Street (and city information) maps Street map programs originated with individual cities. To appeal to business people, vacationers, and locals, information was added about restaurants, hotels, sightseeing, points of interest, airports, etc. Although some of these programs are limited to cities or counties, others provide national coverage. Street map programs are useful for personal and home use, and many have increased in sophistication to appeal to business users as well.

To keep current, the information used in sophisticated programs is updated by CD-ROMs, digital video discs (DVDs), and the Internet. For example, annual updates of white or yellow pages are available on disc, and symbols or icons on some street maps can be linked to Web sites. This means that the user can simply click certain symbols on a map and obtain current information from the Internet. For example, a user can click on the symbol for

- Hotels—for amenities, room rates, and availability
- Restaurants—for menus, prices, and take-out orders
- Theaters—for show titles, starting times, and prices
- Museums—for exhibit dates, descriptions, hours, and prices
- Sport arenas and stadiums—for teams, game times, availability, and pricing

Automobile navigation Mapping and informational software, similar to the programs previously discussed, is being incorporated into automobiles. A small monitor on the dashboard can route the driver to a destination after he or she simply enters an address. A text description can also be provided on the monitor or dashboard, and voice directions may accompany or replace visual devices. These methods will let the driver know when and where to turn. The location of the car can be monitored to show its changing location on the map during the trip, how close it is to the destination, and if the driver took a wrong turn. It can also help locate various places and information about them, as mentioned previously. One of the first commercial experiments with navigational software was conducted in 1994 and involved a fleet of about 50 Avis rental cars in San Jose, California.

Mapping equipment is not only useful to the occupant of a car. If the car is stolen, authorities can quickly find it; if the car breaks down, repair services can be dispatched. Someday a device, possibly included in a PDA (personal digital assistant) or key chain, may be available to help you locate your car if you forget where you parked it at a regional mall or large parking structures.

Similar navigation systems can be provided on a portable computer. If you want to inspect the latest house listings, your laptop can show you how to get to the properties and provide information on each. If you have business appointments, your laptop can direct you to the place of your next meeting.

Certain functions are being integrated among navigational devices, mapping programs, city informational directories, personal address books, and phones. If you need to contact the listing broker with a question on a house or a business client concerning an appointment, a phone call from your cellular phone becomes nearly automatic. With to-do lists, names, addresses, and phone numbers, some programs have the potential to be used along with cellular phones. As of 1998 this is all possible on a laptop computer. Expect cars to better accommodate portable computing devices for this type of use in the future.

Information programs suitable for mapping In part, the impetus needed to drive mapping demand are specific uses and applications. On a mass-market basis, popular informational products are making mapping more useful and important. The two products described below are now often packaged with street map programs and are helping spur the sale of mapping programs.

Telephone directories. A number of phone directories are available on CD and via the Internet. Do you need to find the address and phone number of a lost friend or someone who recently became ill, engaged, gave birth, etc.? Do you want to locate the homes of everyone in your team, club, alumni group, etc.? Phone directories covering the white page listings for millions of individuals across the country can do this. The yellow page directories are especially useful for business and contain information on millions of companies. If you repair or remodel properties and need to know the closest hardware store, lumber yard, nursery, etc., information on each company is available. Commercial directories can also be helpful for personal use, to locate local retailers, service firms, and restaurants. When used with mapping software, phone directories can be helpful for personal use and powerful as business tools.

Address books, contact managers, and PIMs. The marketplace is replete with software products to help us keep track of phone numbers, addresses, and related data for clients and friends. Programs include fairly simple address books as well as personal information managers (PIMs) that often contain a calendar, task list, agenda, the ability to dial phone numbers, etc. Contact managers are upscale address books, or PIMs, which are especially useful for people in sales. Mapping software can be used with these contact managers to plot addresses on a map anywhere in the country. This type of functionality could eventually be incorporated into more sophisticated contact management software as well.

Home Mapping

Home Mapping has various applications, including travel, education (geography), and games.

Travel (trip planners) Travel products are usually nationally oriented with limited city information. They may contain street maps of cities, sometimes as an add-on option, which may include detailed tourist information. Travel software can plan your trip between two points, like the auto club's Trip-Ticks which finds directions and distances between cities or locations, provides information about interesting points along the way, and prints an itinerary with directions, timing, and even costs. Some of these trip planners are GPS capable (determining position via satellite signals), which allows them to show your location on a laptop computer as you make your trip. Thus they become an automobile navigation device.

From a home use perspective, the entire family can become involved in planning a trip and monitoring it. Children may find such programs fun and interesting. They can provide information to help children learn more and may even keep them occupied while traveling.

Trip planners can also be used for business and personal mapping. Although they are most useful for longer trips between states, or at least between cities, trip planners are similar to automobile navigation software and may function as such. Many people involved in sales spend a great deal of time traveling by car. Especially for those managing new sales territories, trip planners can help business people get where they are going and even provide some entertainment and education along the way.

Education (geography) We all had to learn to identify states and their capitals in grade school, but today children can use mapping software that teaches them much more than names and locations. Programs like US and World Atlas can improve a child's ability to relate to maps, which may profoundly affect business in the future. They may also help parents become better acquainted with mapping and its capabilities for obtaining information and data. Although most basic geographic products are for educational purposes and not directly related to real estate and business, they could have a significant influence in preparing the younger generation for using GIS products in the future.

Games Several games have a geographical and mapping focus. One popular CD-ROM game, "Where in the World is Carmen Sandiego?" tests the player's knowledge of geography and proves that geography can, in fact, be challenging and interesting. Another game, "SimCity 2000," combines geographic and real estate issues in challenging "real-world" situations.

Specialization

Specialization of GIS and mapping may occur through customization by the user (or someone assisting the user) or it may be built into the product by a value-added reseller (VAR). Sometimes specialization is brought about by a combination of both. A highly specialized built-in program may have an abundance of data and numerous formulas, scripts (programming), reports, design layouts, etc. Higher-level GIS programs are the most specialized, and those in mapping are the least specialized. Specialization in mapping products is often built-in or bundled with phone directories, address books, or contact managers. In contrast, specialization in GIS often requires more user-specific customization.

All the major GIS software vendors offer scripting languages for programing customized applications, e.g., MapBasic from MapInfo, Avenue from ESRI. Some vendors have begun unbundling GIS functions so that a programmer can buy a much-reduced set of GIS tools for building simpler, cheaper, more focused applications; MapObjects from ESRI is a prime example. An added benefit is the compatibility of these tool sets with widely used graphical user interface (GUI) development tools such as Visual Basic and the ability for some to develop Internet applications.

Increasing Built-in Specialization

Traditionally, most GIS programs have generic appeal to a wide array of academic, government, and business users. Throughout the 1980s and the mid-1990s, there have been relatively few GIS programs with built-in specialization for real estate. It is largely up to the user to customize a GIS program for his or her needs either in-house or with outside assistance. This may be appropriate for large organizations, but it is very daunting for smaller companies.

To enlarge their customer base to include vast numbers of medium-sized and small businesses, GIS software and data companies are finding it necessary to develop more programs with built-in specialization. Some of these programs for real estate professionals have been developed within software companies, but others are created in partnership with data and demographic companies, value-added resellers, or system integrators (SIs).

Using GIS and Mapping to Generate Internet Maps

The extent of specialization of a GIS or mapping system may relate to its use as a publishing engine to generate maps. This is becoming especially important for products that provide certain end products over the Internet and intranets. The final products from these GIS or mapping engines may include a map showing a route from one place to another along with specific directions. A demographic analysis may include a map with concentric circles around a point and an accompanying report. A multiple listing service can generate a map indicating listings and/or sales. A research report can show a map indicating the submarkets of a metro area along with vacancy, rental, and other data. Such publishing engines usually provide little opportunity for the end user to modify the final map. It is possible, however, for the user to move the map into a report or a letter and then annotate it.

The use of GIS and mapping on the Internet is exploding. GIS and mapping allows extensive use of maps by customers over the Internet, and intranets allow more internal and private corporate use of maps. For example, VISA has an Internet site that allows individuals to type their address and find the closest ATM machine. A real estate professional could use a similar function to find the nearest listings or sales from a particular address, the closest restaurant, store, self-storage center, etc. Using intranets within a company, it is possible to access and map information such as the location of any employee or regional and local offices, sale performance by office, number of properties sold, square footage sold or managed by region, etc. Most important, these functions can be easily accessed throughout the country or even the world.

Leaders in providing sophisticated GIS tools for the Internet include ESRI, Intergraph, MapInfo, and AutoDesk. Those offering cut-and-paste, simplistic maps on the Internet can be found at www.mapquest.com, www.vicinity.com, and www.mapsonus.com and by using popular search engines such as Yahoo! and Excite.

Integrating GIS and Mapping into Other Programs

Another important and escalating trend involving specialization is the incorporation of GIS and mapping into other products. This feature may be offered in programs that operate on a stand-alone PC, but increasingly seems to involve networking and Internet/intranet products. The types of real estate programs that lend themselves to mapping and GIS integration involve asset management, brokerage (sales), leasing, and property management. Such programs use components of mapping or GIS programs to generate maps. Hence, in this specialized area, mapping or GIS is used as part of another program, not as an overall solution.

The relatively new trend of incorporating mapping into other software began in the late 1990s and indicates a change in the direction of GIS development and marketing from only a few years before. In the early 1990s, GIS was marketed as an overall solution to many services and industries, including real estate. Although GIS may still be an end-all solution in many situations, more applications will arise for GIS and mapping as a part of other programs. This combination could become so seamless and simple that users might not be aware of which specific GIS or mapping product is in use, and they may not care. Also, the concepts and terms of GIS will become more refined and easier to understand.

Summary and Conclusions of GIS and Mapping Software

Challenges for Growth

Although a GIS product could be readily available and easy to learn for many specific real estate services, the costs to develop and market such a product on speculation can be prohibitive. Often the perceived risks of developing customized GIS products do not justify the perceived rewards. This has been the case for many complex GIS solutions where the markets are so specialized and the number of potential customers so low that revenue forecasts have been unattractive. Only when GIS and mapping products provide solutions that can make money or generate savings that exceed their cost many times over will they become more commonplace in real estate.

In comparison to developments in forestry, oil and gas, transportation, insurance, and marketing areas, GIS and mapping has developed relatively slowly in real estate. These other industries have GIS applications that are more easily justified in terms of cost.

One explanation for the dearth of GIS and mapping in real estate is attributable to the type of people in the industry. Many real estate professionals and practitioners have little awareness of GIS, a relatively low technology level, and a reluctance to use something new and unproven that requires significant cost and time. However, this is expected to gradually change as more technologically proficient people enter the real estate field.

Another reason GIS may not be readily adopted is that it can threaten real estate incentive fees and commissions. Many real estate executives have incentive positions in acquisitions/dispositions, management, brokerage, or leasing. If GIS provides a more objective view that dampens deal making, it would not be warmly received by commission-based real estate executives.

Also, GIS, in contrast to mapping, is inherently complex since it involves greater analytical, database, and multivariable capabilities. More complex solutions limit potential users and the size of the market. Further, more complex solutions often involve more customization to a very specific service or task. Hence, markets such as real estate create challenges for product development.

Favorable Demand Factors

More impetus for using GIS in real estate is sure to develop. Some experts predict that demand will arise via public authorities. If assessor, planning, and zoning offices provide access to their files through a GIS system and mapping interface, more people will expect this information. Public agencies may also make it much easier for businesses to obtain the data required.

Although these innovations will have some impact, the demand for GIS in real estate will probably occur when a few firms begin to implement it and laud its benefits. Traditionally, professionals in a given industry adopt a herd mentality. Everyone wants to be in the hot markets, not the bad markets. Everyone wants to be involved with a hot property type and to avoid those that are unpopular. As more real estate firms use GIS and mapping, they may recognize that advantages can be obtained.

Another force driving the demand for GIS in real estate may be mapping products. Such products are very affordable and easy to use, and they provide some straightforward applications and solutions for real estate professionals. Mapping products are an excellent introduction to GIS.

As larger real estate organizations become more automated, GIS will emerge as a means to facilitate better understanding of internal information. Customization with mid- and high-level GIS can provide a wide range of industry and professional solutions for problems involving property operations, accounting, property management, insurance, taxes, evaluations, and engineering studies. Large-scale asset management systems with object-oriented capabilities are expected to incorporate mapping or GIS. Thus, additional demand will be generated in various segments of real estate and related industries.

GIS and Mapping Trends

In summary, the key to the growth of GIS and mapping in real estate lies in the combination of several trends:

- Development of products with built-in specialization, including the bundling of data.
- Integration of GIS and mapping into other software products.
- Use of GIS tools on the Internet and intranets.
- The "herd" factor—with more real estate firms using GIS, an increasing number of firms will decide they need to acquire GIS as well.

Perhaps the most significant trend is the adoption of mapping as legitimate business tools as evidenced by major software companies such as Microsoft.

DATA

Data Sources and Types

Generally speaking, data sources are either primary or secondary. Primary data are information gathered directly, including internal information. Secondary data are gathered from another source. Data may also be either public or private and available either free or for a fee. Although public data are often thought of as free and private data available for a fee, sometimes the reverse can be true.

One method of organizing data is to consider the source. Four major sources are

1. Academics—including colleges, universities, research centers, and foundations
2. Government agencies—local, state, and federal
3. Associations—trade organizations and nonprofits
4. For-profit companies

Whatever the classification, all data for GIS and mapping purposes are referenced geographically. An estimated 80% of all computerized information has some type of geographic reference, such as a street address. This indicates the tremendous potential of GIS and mapping for use in integrating all needed data sources, not only for real estate, but for numerous other applications as well.

A key issue is whether the data are proprietary, in which case copyright restrictions will often limit their use. Restrictions apply not only to data from profit and nonprofit groups but also to some public data.

Many local authorities are wrestling with the accessibility of their data. Data can be provided free of charge, at a minimal cost to defray administrative expenses (possibly including GIS), or for a fee that provides some profit to offset other costs within the local authority. Related questions include: How public should public data be? Should the price you paid for your home be readily available to anyone? At present, the verification of sales data and other information varies considerably and is usually determined by each state. Hence, the success of some data vendors and GIS or mapping systems will probably vary a great deal from location to location.

The issue is whether the disclosure of certain information between individuals and other entities is necessary to create more financially sound and economically efficient markets that can benefit the general public. For example, could the problems associated with the savings and loan crisis of the 1980s, which cost taxpayers hundreds of billions of dollars, have been avoided by increased disclosure of real estate transactions?

As the old computer saying goes, “Garbage in, garbage out (GIGO).” Even if you use the best software and hardware available, the results will be meaningless if improper data and analysis techniques are applied or if the wrong types of data are used. Since real estate is a multidisciplined field, the data requirements can be extensive. The several types of data applicable to GIS in real estate are examined here, and considerations pertaining to the quality of data are discussed.

Data Classifications

Digital Data

Digital files (also called *graphics files* or *boundary files*) are the essential ingredients in making a map. They consist of *x,y* coordinates (usually longitude and latitude) that define all the features on the map. A feature defined by

- A single *x,y* coordinate, such as a house, is called a *point, place,* or *location.*
- A string of *x,y* coordinates, such as a street, is called a *line.*
- A string of *x,y* coordinates that loops back and closes on itself like the outline of a zip code zone or a county, is called an *area* or *polygon.*

Information that can be linked to one of the above digital features is called *attribute data*. For example, if a home is represented as a point on the map, its attribute data can include the home's age, size, asking price, number of bedrooms, etc.

Care should be taken to acquire digital files that

- Are compatible with the particular GIS or mapping program being used
- Align correctly with one another, e.g., streets with census tracts
- Fulfill the needed degree of accuracy, visual appeal, and resolution
- Are updated at acceptable intrevals
- Are tailored to the data being used, e.g., demographic data organized by zip code zones cannot be used with census tracts

Each of the three types of digital files—points, lines, and areas—will be briefly discussed.

Points These files provide the names and often associated data for property types, businesses, public buildings, churches, parks, schools, hospitals, etc. These files may be available locally, regionally, or nationally.

Databases for different types of geographic point files are available. One such database, known as GNIS, includes the names from United States Geological Survey (USGS) maps. Business establishment data, with standard industry classification (SIC) codes, staff size, sales volume, etc., are available from Dunn and Bradstreet and many demographic vendors. National data are available by property type as well, especially those focusing on shopping centers and restaurants. Shopping Center Profile, provided by the National Research Bureau, covers thousands of shopping centers and includes associated data. Other databases identify restaurants, hotels/motels, airports, golf courses, colleges, etc. These files are available on CD-ROM. Increasingly, however, mapping and GIS software, including Internet sites, incorporate such data.

Lines The most frequently used line files are street center line files, but other line files frequently include railroads, utility easements, rivers, bodies of water, etc. The U.S. Census Bureau's Topologically Integrated Geographic Encoding and Referencing (TIGER) system has made such files more accessible on a national basis. Replacing the older DIME system, TIGER combines 1990 census data with USGS geographic coordinates. The original 1990 TIGER files have been revised in 1992 and 1994 TIGER releases. This revision improves address matching from an average of about 40% to nearly 90%. Although the street files you acquire should reflect the most recent revision, many vendors have significantly enhanced the TIGER files. A vendor product may be particularly valuable to those dealing with a recently developed area. TIGER files identify most, but not all, streets in the nation. However, newer streets are usually missing, necessitating occasional data updates. As an alternative to maintaining current street and geographic files, users may be able to update the TIGER files themselves with proper equipment and software.

One of the largest enhancers of street files is ETAK. In addition to refining the accuracy of TIGER files, ETAK improves the color shading and width of street files and adds certain landmarks. Geographic Data Technology (GDT) and Business Location Research (BLR) also offer specialized street files. Both BLR and GDT have several advanced street address databases that provide more complete address-range information. BLR and GDT also offer traffic volumes for major roads in more than 100 cities.

All street files are not equal. Some files have

- Better address matching
- More recent streets and landmarks
- Differentiation of street size by color, e.g., secondary, primary, state route, interstate, etc.
- The ability to easily turn on and off different levels of streets in a GIS presentation

Street files can be purchased for the entire country, by state, by county, and sometimes by city. Because these files often require extensive computer memory (typically one megabyte for a small city), they are often provided on CD-ROM.

Areas For the U.S., the basic geographic boundary files include:

- States
- Counties, including census county divisions (CCDs) and minor civil divisions (MCDs)
- Metropolitan statistical areas (MSAs)
- City boundaries
- Census tracts
- Census block groups and individual blocks
- Postal area zip codes (three-digit, five-digit, and zip plus four)
- School districts

Some real estate data vendors provide their submarket boundaries in GIS-compatible form. Some of the products provided by these vendors cover environmental hazards, flood zones, earthquake fault lines, crime areas, school districts, Superfund sites, hurricane-prone areas, etc. A few of these vendors are On Target Mapping, Environmental Risk Information and Imaging Services (ERIIS), and Environmental Data Resources (EDR).

Primary geographic areas are used by most real estate professionals. Others areas that are available in digital form but are not likely to be used include areas of dominant influence (ADIs) and designated market areas (DMAs), congressional district boundaries, and cable TV franchise areas.

Digital files are readily available through mapping firms, demographic vendors, and third-party data vendors. They can be purchased for the entire country or by state, county, or city. Because digital area files often require much storage space, they are usually provided on CD-ROM.

Comments

In addition to using data from the national sources mentioned, a GIS program should allow geocoding so that if location coordinates are available, a large number of specific points, locations, or properties can be quickly identified. An increasing number of data vendors are offering geo-coordinates for property sales, rentals, and listings. Of course, individual labeling can be accomplished with most programs, but this requires more time and can be less accurate. Sources on the Internet also allow users to obtain coordinates simply by providing an address.

When selecting boundaries files, keep in mind that they become dated. Zip code areas can change as frequently as every month, and the data can change significantly within a year. MSA and city boundaries also change, but census boundary files are updated only once every decade. Therefore, when acquiring digital files, it may be prudent to learn the cost of updates and the likelihood of the firm delivering those updates on a regular basis.

It is important to match the digital files with the data being used. For example, if you use 1990 zip code data with 1993 zip code boundaries, you will have many zip codes without any data. Take care not to overlook this issue, especially when purchasing programs bundled with boundary files.

If you plan on using a different mapping program in the future, it may be difficult or impractical to transfer or convert the boundaries to the new program. To acquire new digital files or convert them to a new format could cost thousands of dollars. Even if you obtain new files, they could be inconsistent with maps created by the old program, especially if they are from a different vendor. Try to anticipate these situations to minimize problems down the line.

The pricing of digital files varies considerably depending on the areas covered, the enhancement techniques used, and their currency and resolution. The suppliers of these data are expected to change, however, as these data are incorporated into more programs. The need to wrestle with accurate files should diminish for many end users. Costs have dropped dramatically in recent years and should continue to fall.

Demographic Data

Demographics is the science of vital social statistics. Most U.S. demographic data are based on census data. Census demographic data are normally provided from two sets of information. The first set, STF1 (Summary Tape File 1), corresponds to the short census form and reflects population, income, age, sex, ethnicity, and households. The second set, STF3, provides data from the long census form and includes detailed housing, education, income, and employment data. A letter after the set number reflects the boundary file associated with the data. For example, in STF3A and STF3B, the "A" indicates census tracts and the "B" indicates zip codes.

A complete census survey is made available every 10 years, and these data are now on CD-ROM at very affordable prices. Pending changes in census techniques call for conducting the census more frequently but gathering less extensive data.

Although demographic data can be acquired directly from the government, it is often easier to use the data supplied by a third-party vendor who has also provided the boundary files for the mapping software used. The procedures for transferring data into a GIS program can also be facilitated through the software built into certain demographic data products. The software may allow for the selection of geographic boundary areas (or the creation of custom areas), the selection of multiple variables (which can be saved as lists), and linkage to mapping software.

As census data age, it becomes more important to obtain demographic estimates and forecasts. Updated information for local areas can be acquired from many planning agencies, but for forecasts on a national level the most convenient source is a demographic vendor. In some areas, especially California, demographics can change fairly rapidly. Census data that are several years old may be inappropriate, especially if a region has experienced much growth. To obtain updated demographic data, a user can choose from among several major vendors. These vendors provide comprehensive data down to the census block group. Other vendors also provide econometric data, but their demographic data are usually limited to the county level. The quality of vendor forecasts can vary depending on the procedures employed.

The principal delivery method for third-party data providers is CD-ROM, but Internet and other alternatives are also available. With your computer linked into a data provider's computer through a modem, you can access and download specific data as needed. Many government agencies also have a wealth of data available on-line.

The pricing structure for demographic data varies depending on the type of census data (STF1 versus STF3) and historic versus forecast data. Forecast data can be purchased for selected time intervals, areas (local to national), and update frequencies. The cost may be a set charge for unlimited usage of an on-line service or an amount based on the quantity and type of data used. A premium is charged for updated and forecast data.

Like digital data, demographic data can be purchased not only from data vendors but also through GIS software and other vendors.

Economic Data

Economic analysis is concerned with the production, distribution, and consumption of goods and services and focuses on employment patterns among various types of industries. Economic data are essentially of two types: employment data that relate standard industrial classification (SIC) codes to geographic areas and business-specific data (e.g., address, contact person, phone number, SIC code, sales, employees) that relate to geographic areas. Employment and business data change constantly, are updated frequently, and can often be expensive.

Employment data Employment data are provided by different government agencies and reported monthly by either place of work or residence. Although detailed employment data are available directly from the government, forecast data are only available from major economic vendors such as Regional Financial Associates (RFA), Dodge Resources Inc. (DRI), and WEFA. Such firms profile economic trends for the nation and by region and city. In particular, they supply historical and forecast employment data for major and minor SIC codes that can be easily mapped by MSA or at the county level. Data are available in hard copy and most computer formats. Pricing begins at more than $1,000. Small forecasting groups can provide similar employment forecasts, but they do not offer as wide a range of services and their data are not as detailed.

Business data Business data cover business formations (starts), business failures, job ads, and information about existing businesses. One of the leaders in this area is Dunn and Bradstreet (D&B). The company has assembled a database from approximately 10 million U.S. firms to provide information on trends in business failures and new incorporations. D&B produces credit reports on businesses, which can be particularly useful when prospecting for new clients. D&B data are often purchased by third-party vendors who modify them to offer additional services and enhanced data.

Real Estate Market Data

Just as the real estate market is fragmented by location, property type, and services, data pertaining to real estate are fragmented. This discussion highlights major groups of data and national sources.

Supply and demand Supply data include information on vacancies, building permits issued, buildings under construction, and proposed buildings as well as the total land area zoned and available for development.

Demand data consist of two sets of variables: independent and dependent. Independent variables such as employment and population growth and household formations explain dependent variables such as absorption, rent, and price.

Both supply and demand data can be viewed on an aggregate basis, as a percent of change, or as an index. Examining one without the other can be misleading. For example, a market with a 5% increase in supply may not seem significant. However, if the market is large, say 50 million square feet, and this 2.5 million-square-foot increase is primarily located in a single submarket, it can be very significant.

The following sources present supply and demand data on a national basis. F.W. Dodge provides information on historical and forecast building construction as does its affiliated company, the Real Estate Analysis and Planning System (REAPS). REIS Reports offers information on vacancies, rents, property data, and new properties for apartment, office, industrial, and retail properties in 50 to 75 markets. These two services encompass metropolitan, submarket, and even property-specific data and are largely oriented towards national investors and the securitization of large portfolios. A research division of the brokerage firm CB/Richard Ellis, known as Torto/Wheaton, obtains information from its local offices covering some 50 different markets for industrial and office properties and furnishes some data on apartments as well. Landauer Associates R-E-Search provides similar information on-line. Information for both is available on a submarket basis. All offer some degree of GIS support—e.g., Torto/Wheaton provides its submarket boundaries in a GIS format.

There are other national sources, but much information is in transition from print to a digital format for use with GIS. Moreover, national supply and demand data are limited and may focus on only one or a few property types. Among the many local data sources, some provide computerized data, but sources can differ in their consistency, frequency, accuracy, and methodology.

Performance Performance data can include information on changes in prices, rents, net incomes, rates of return, capitalization rates, and gross income multiples. The sources of supply and demand data noted above may also provide some performance data, as does the National Real Estate Index, which was developed by the Liquidity Fund and is now part of CB/Richard Ellis.

Two major commercial property indexes provide returns from appreciation (derived from appraisals) and income (from actual operating performances). One index is maintained by Evaluation Associates, Inc. of Norwalk, Connecticut, and the other is by the National Council of Real Estate Investment Fiduciaries (NCREIF), formerly cosponsored by the Russell Company of Tacoma, Washington. These indexes are published quarterly and cover several investment property types. The (Russell)NCREIF Index was the first index available in computer format. Historically, this information has only been available to academics and NCREIF members. Although these indexes cover a large number of national markets, the size of the property base has not been sufficient to provide much submarket data.

Property and transaction Public records contain a wealth of information from official departments such as the tax assessor, clerk of records, zoning and land use authority, public utilities, and building permit office. These sources can provide a great deal of property-level detail and sales data as well as one of the most comprehensive sets of maps imaginable—assessor's parcel maps. Unfortunately, in many if not most areas, these public data are unavailable in computer format. However, several national firms and numerous regional firms assemble this public information, repackage it, and supply computerized property and transaction data. Market leaders include DataQuick, Experian, Transamerica Intellitech, and COMPS, Inc. Others can be found via their trade association, the Real Estate Information Providers Association (REIPA).

Data available by property include property address, owner, loan and default data, building characteristics, lot size, legal description, assessment and zoning information, and flood zone data. The quality of the data can vary tremendously. Information for some properties may be incomplete or even erroneous. Accuracy depends on the thoroughness of the local government entity as well as the firm repackaging the data.

The data from these vendors can be used for a variety of real estate purposes. Appraisal, investment and brokerage are common applications. One might also employ property information for market prospecting, creating mailing lists and labels for mortgage prospects, loan collection, and even auto-prospecting.

Many of the national vendors noted above supply their data on-line and eventually most, if not all, are expected to do so. Many also provide it on CD-ROM and offer floppy disks. The ease with which these data can be transferred into a mapping program varies by data vendor. Other methods of providing data—e.g., on microfiche, in hard copy form, by fax service—are not compatible with mapping programs. The frequency, quantity, delivery method, and types of services all affect the price of the data.

As current and relevant data become more available, the importance of GIS and mapping to process and analyze the data will increase.

Aerials, Photos, Diagrams, and Maps

Aerials, photos, diagrams, maps, and other types of images can be incorporated in a GIS or mapping system. Images can be scanned and appear in raster format, allowing them to be used as a backdrop to vector-based files such as street maps. Besides functioning as a map layer, images can also serve informational purposes. For example, a symbol depicting a property can be activated by a click of the mouse to show the property's survey map, site plan with improvements, floor plans, architectural plans, topographic and soil conditions, etc.

Aerial images can be obtained from either a local aerial photography vendor or a satellite image company. Satellite images are obtainable from several sources including Spot Image Corporation and Space Imaging/EOSAT, both of which provide worldwide coverage. These firms also sell their images to data and map vendors, who then either resell them or incorporate them into mapping programs. To showcase Windows NT's capabilities, Microsoft introduced an enormous database containing satellite imagery freely available to the public. Some of the files are so large, however, it can be awkward to download these images. Soon satellite imagery will become available at a resolution of one meter—i.e., the images will show everything that is one meter or larger in size. The prior resolution standard was five to ten meters. Low resolution imagery is available at lower prices than higher resolution imagery.

Just as satellite images are commonly imported into GIS, standard photos can be used as well. Photographs can be cropped, sized, and attached to points within a GIS program. Photos can be imported using a scanner or a digital camera, available in either color or black and white. The technology of digital cameras holds tremendous potential, not only for business applications but eventually for consumer use as well. An extension of the photograph, video images can also be imported into most programs, but the process may require special software and hardware.

Detailed layouts of buildings and floor plans are useful to those managing the facilities of large corporations, institutions, or government agencies, but oversized blueprints and plans are often awkward to store. Blueprints and plans can be imported into a GIS by scanning them or tracing the images on a digitizer.

Often users need to incorporate special maps along with data. Some assessor offices offer digitized block and parcel tax maps that can be useful in a variety of ways. Zoning and land-use maps also provide an excellent base over which other information may be viewed. Submarket areas, voting districts, sales territories, etc. can be created by using existing software to aggregate various digital files (such as census tracts or zip code zones) into new files.

The integration of different types of images into a GIS or mapping program is costly, time-consuming, and sometimes complicated, but it can consolidate items that are often misplaced and hard to find or store. The time spent assembling diagrams, photos, and maps can be reduced and allocated more productively.

The future of GIS or mapping in real estate depends on the development of appropriate and meaningful databases, both locally and nationally. As local data sources are assembled or developed by public agencies and national vendors, the real estate industry will ultimately benefit. The result could be an enormously extensive national database encompassing all the variables needed for more sound decision making.

Supplemental Sources

Many additional data sources can be made available, especially on a local basis, to provide a level of understanding and analysis that was previously inconceivable. For example, analysts can gather data on:

- Major employers and employment centers
- Public transportation routes (e.g., bus, trolley, subway)
- Crime districts and statistics
- Environmental concerns:
    - Air quality
    - Water quality
    - Oil types
    - Earthquake zones and faults
    - Underground toxins
    - Superfund sites
    - Aircraft noise
    - Areas of endangered species
    - Disease incidence
- School districts and schools
- Significant places:
    - Hospitals
    - Churches
    - Cemeteries
    - Restaurants
    - Theaters
    - Parks
    - Golf courses and tennis courts
    - Public buildings (e.g., police and fire stations)
    - Ports and docks, airports, train stations, bus stations
    - Military bases

Source: Schott Consulting Services

In-House Information

Although in-house data sources are presented last in this discussion, they are often the first to be considered. However, some words of caution are needed. The most knowledgeable person about what data can be used is the user and his or her coworkers. A needs assessment study can be performed by a consultant or system integrator, but the most pertinent information lies somewhere within your firm—be it in file cabinets, on desktops, or with subcontractors. The more scattered the data are, the more help you will probably need in pulling it together.

The most obvious source of data for GIS are the properties with which you are most actively involved. These can be properties that are listings, under management, being leased, in a portfolio, under foreclosure, being auctioned, or being developed. Properties with which you are less directly involved may also serve as excellent sources of information. They could be properties targeted for prospecting, properties involved in prior assignments or sales, properties with loan defaults or those involved in negotiations, competitive peer properties, comparable rentals, etc.

Do not forget the administrative side of your business, especially if you work in a larger company. How about creating a map showing your headquarters, regional offices, smaller offices, or corporate holdings? What about producing a map of the territories for each region?

Also, remember to consult the marketing people. They usually have large databases of existing, prospective, and past clients. These files often can be transferred into a GIS or mapping system with relative ease, especially when marketing territories are clearly defined or only one person has marketing responsibilities.

Think about the benefits of bringing these different subsets of data together in a GIS. One might see new relationships and discover new insights. Numerous decisions can be affected, such as plans to close, open, consolidate, or relocate offices.

The information used in a GIS system for real estate need not be limited to real estate; it can be used for other functions of the firm. For example, locating the residences of all company employees might be of interest to the personnel or benefits department or to the CEO, especially if a new office location is being considered. If conferences, seminars, or meeting functions are held periodically, potential sites could be added to a GIS and compared to the locations of the attendees to find the most convenient site. Though the focus of this writing is on GIS in real estate, it is important to remember that GIS has other applications as well.

Final Comments on Data Issues

Just as books have reference notes, maps should indicate the data sources used. Although often overlooked, footnoting maps offers numerous benefits:

- You can easily identify the most frequently used sources and focus on the most important ones.
- If you are reviewing the work of someone else, you might need to know what data sources and boundary files were used.
- Only by checking references can you discern whether maps were prepared with recent or dated files.

Internet sites are easily accessed through browsers or the built-in operating systems of some programs. It is expected that these browsers will evolve with Web-casting capabilities for real estate specific applications and allow for an automatic retrieval system (using push technology) in desired formats. This not only offers the ability to collect new data and update programs when available, but also to obtain other real estate information such as city and market reviews, news, regulatory changes, etc. This information could then be utilized in a separate analytical package or one that may be built into the Web-casting program. Thus Web-casters could evolve into a total real estate information solution, of which real estate data and GIS/mapping software would be a closely integrated part.

On-line data will be the preferred delivery method, especially with Web-casting programs. However, other delivery mechanisms, such as CD-ROM or digital video/versatile discs (DVD), may be more appropriate for extremely large databases and archives.

With so many real estate data sources available, it may be difficult to know what information you need and where to get it. The types of data carried by these vendors are changing and vendors may come and go. Moreover, mistakes are costly. Accessing the same data from different sources (and sometimes from the same source) can make a fee jump from a few dollars to a few thousand dollars. These perils are being resolved by those who are actively involved in making the future a reality today; other users are content to sit and wait. Those in the vanguard of change may or may not reap more benefits than those who are patiently awaiting new developments. One thing is certain though, change is inevitable. Those who wait too long to adapt to change will face obsolescence and the demise of careers and companies.

MANAGEMENT ISSUES

For a GIS or mapping system to be financially justified, the tangible and intangible returns must outweigh the costs. Since GIS has multiple applications, it is wise to implement those features that provide the greatest financial benefits first. Because the expense of mapping is nominal—a few hundred dollars at most—a cost-benefit analysis is unnecessary. Therefore, this section will focus on GIS and cover both limited and sophisticated applications.

Tangible Returns

Four measures of tangible returns are discussed here: increased business, increased productivity, new services, and reduced costs.

Increased Business

The most important reason to install GIS may be to increase business. One commercial broker using GIS has attributed 50% of his new business to the use of GIS.[1] If employed properly, GIS can provide your clients with a better understanding of important issues. A GIS can produce impressive presentation materials for existing or potential clients, and its use conveys a heightened sense of knowledge and professionalism. Although a GIS requires additional capital expense, it can help a firm provide new services and thereby augment revenues. New income is complemented by a reduction in costs over time due, in part, to the increased productivity GIS can facilitate.

Increased Productivity

Once the learning curve is overcome, it is possible to accomplish more work in less time with GIS. Less time will be needed to

- Prepare presentations
- Create maps
- Produce demographic reports
- Locate listings, sales, or assets on a map
- Determine the best route(s) to take when visiting a property or a group of properties.

For example, when photos are required in a report, the preparer will no longer need to search for the negatives, find the specific photos desired, have the selected negative printed at the photo store, pick up the photos, and paste the photos on each set of reports. An integrated GIS product that can accommodate images from digital cameras or scanners eliminates most of these functions. GIS can also facilitate the loading of data into spreadsheets and graphics programs. By allowing individuals to perform tasks more quickly, GIS will reduce certain expenses and enable employees to take on additional responsibilities.

New Services

With GIS and mapping, firms can offer new services without incurring much additional expense. As mentioned earlier, some brokers use mapping to provide their clients with maps identifying listings

1. Gary Beban, from the keynote address at GIS in Business '94 Conference in San Francisco, June 1994.

and even sales with property addresses and summary data. Real estate appraisers and consultants using mapping or GIS can now create neighborhood maps that show schools, shopping, bus routes, employment locations, and other pertinent data that previously could only be described verbally. With GIS, this information can then be spatially related as part of an improved analysis. In addition, traditional paper reports can be augmented with demographic data, maps, images, charts, and even videos and be transformed into a multimedia experience.

The Internet is also advancing our use of mapping and GIS. In 1996 it was noteworthy that some Web sites could identify an address on a map. About a year later, directions between two points became available, along with maps showing each turn. More sophisticated mapping and GIS applications will become increasingly common on the Web.

Reduced Costs

Time is money and GIS can save time. It also can reduce, if not eliminate, outsourcing, which provides direct savings. For example, one brokerage firm spent more than $100 each for maps and demographic reports that were periodically ordered by approximately 100 companies. When this firm installed a specialized demographic mapping system, annual savings of over $30,000 were realized.[2]

Intangible Returns

Some of the most significant benefits of GIS cannot be readily quantified. Some of these intangibles are discussed below.

Improved Analytic Capabilities

The ability to use layered data and visualize them spatially can provide a user with greater insight and understanding. Relationships among specific properties, services, and various other factors can be analyzed in new ways. For instance, the effect on property value attributable to a property's location in one school or tax district rather than another can be better understood with GIS. GIS helps manage information better and allows comparisons to be more easily drawn. Furthermore, standard procedures can be automated, which permits more timely and frequent analyses.

Loss Avoidance

Perhaps the least recognized benefit of GIS is its ability to help you avoid unprofitable real estate deals. In the heat of competition and the spirit of a fee-driven business, inadequate analyses have caused ruinous results. Using GIS one can analyze a property more thoroughly, which should help identify both good deals and bad deals and promote more intelligent decision-making. In short, GIS can help you protect capital and avoid disastrous results.

Improved Image

When state-of-the-art capabilities improve a firm's performance and lead to better decisions, the regard and stature of the company rises. Companies that can command greater respect can also command higher fees, attract and retain better employees, and create a more profitable firm. Image and respect are not easily measured, but they are tremendously important.

Compliance with Regulations

Some organizations may require that a GIS be used by its suppliers and subcontractors. Just as some lenders require that a particular standard appraisal form or type of appraisal software be used, it appears that the next step could be to require the use of GIS software, which could augment the firm's own GIS system. Such a requirement could apply not only to lending but also to site selection studies, property and asset management, due diligence, acquisitions and dispositions, etc.

2. L. J. Hooker International, Atlanta, Georgia, 1989.

Financial Considerations

The cost of implementing GIS depends on the scale of the system required to satisfy your needs, the time frame you choose, your existing resources, and the assistance you will require. A mapping program used by a broker or appraiser can cost under $100. The costs for a large corporate or government installation with a large number of users can exceed $1,000,000. Since most real estate companies are relatively small, this discussion will focus on financial considerations for a small to medium-scale GIS.

The financial considerations for establishing a GIS consist of start-up and continuing operations costs. The costs cited here relate to a stand-alone system or an additional computer on a network; either provides GIS access for only one user at a time.

Start-up Costs

A few basic categories are described. Additional fees may be charged for preliminary studies such as a needs assessment or to cover the training of in-house staff.

Software. The price for powerful GIS software has dropped dramatically during the 1990s. Nevertheless, at the time of this writing, popular GIS programs bundled with standard boundary, street, and census data files begin at prices of $1,000 to $2,000—much more than word processing, spreadsheet, and other popular software.

The price of a program is largely determined by its features and the category into which it falls. Not surprisingly, the more sophisticated and full-featured the mapping program, the more expensive it becomes. Users need to be aware that some programs have add-on products containing additional features (e.g., program language, runtime modules, conversion programs, Internet map creators, SQL database capabilities) that can be relatively expensive. A program may seem inexpensive, but you must first add up the costs of add-on products, boundary files, and data that are not included with the software. Also, some programs may work well with your existing equipment, while others may require upgrades or new equipment. Hidden cost components are easy to overlook.

Although the bundled software-and-data packages provide a good value, they can often be dated and may contain boundary or street files that are not as accurate as others. If your work involves only one city or state and you require fine detail and accuracy, you may need specialized files. More detailed street files, such as elaborate county street maps with enhanced features and capabilities, cost from a couple hundred dollars to well over a thousand dollars. High-resolution zip code boundary files can be even more expensive. If you need nine-digit zip codes instead of five-digit zips or census groups instead of census tracts, your costs will be higher. The more detailed and recent the data and the broader the coverage, the more costly the digital and attribute data files become.

Data. Data can be obtained from primary and secondary sources. Often primary, or internal, data are thought to be free. This may be true if the data within your company are already assembled so they can be readily used within the software. If not, time and money must be spent to make these data available. There are companies that specialize in organizing and converting internal data for use within companies.

Obtaining secondary data from the government and private sources can be quite expensive and is often the most significant expense associated with GIS. Anywhere from a few hundred dollars to tens of thousands of dollars can be spent for annual demographic, economic, real estate, and other data.

Staffing, assistance, and training. If internal capabilities are not available, outside assistance is required. Although expenses for staffing, assistance, and training may not be allocated to the GIS installation, they must be considered nevertheless. For example, an Information Services (IS) staff person may set up a GIS for a group within your firm, but those charges are general corporate overhead and may not be allocated directly to the group. Consulting fees may be negligible or they may run up to thousands of dollars. Training, including travel, could vary between a couple hundred dollars and a couple thousand dollars for each user. The total cost is also influenced by the number of people involved.

Hardware. A good system unit for GIS today begins at $2,000 but can easily cost twice as much with increased storage capacity, monitor size, quality, and so forth. The most expensive peripheral could be the printer. Ink jet printers sell for a couple hundred dollars, but thermal dye transfer printers may cost a few thousand dollars and the plotters required to print oversized maps can cost even more. If existing computer equipment is available, hardware expenses may be limited to the cost of upgrades.

Total initial costs. Start-up costs for GIS vary tremendously. For someone who uses his or her own equipment and data and does not require additional assistance or training, a powerful GIS program can now be obtained for several hundred dollars; more typically, GIS software costs $1,000 to $2,000. If a scanner, printer, digitizer, and other peripherals are added to the system, an area larger than one desktop is required. This will necessitate not only the purchase of the equipment but also more furniture and possibly more office space. If an additional employee position is created and expensive economic and real estate data are acquired, along with enhanced street and boundary files, a first-year budget of $50,000 to $100,000 may be needed.

Continuing Costs

Once the initial investment in GIS is made, ongoing costs will be incurred. These should run less than the initial costs unless the GIS is continually expanded to include other staff and other departments. Ongoing costs can be expected for the following items:

Demographic information, which historically has been relatively inexpensive shortly after a census or a couple years after a new decade begins. When more time has elapsed since the census, updates and estimates become important, and the data for the current year and forecasts for the next five to 10 years are more expensive. The new census data-gathering procedures expected in the future could affect demographic data costs. New demographic data forecasts may be required more frequently, perhaps every year. This can cost hundreds or thousands of dollars depending on the geographic areas and time sequences needed.

Economic data, if acquired from the government, can be inexpensive, starting at around $100. Fees for private sector economic data services begin at a few hundred dollars and can exceed thousands of dollars. Again, the costs vary depending on the area covered, the time periods needed, the vendor, and the quality of the information.

Digital file updates, with information on streets, zip code zones, municipal boundaries, districts, etc., are usually priced at a percentage of the initial purchase price. A 25% annual surcharge for quarterly updates is common.

Real estate data, consisting of property information, sales, rentals, zoning, etc., change frequently and are usually updated weekly or monthly. The cost of real estate data can be greater than the cost of any other type of data. Local coverage can cost thousands of dollars each year, while annual expenses for national coverage can exceed $20,000 for a single vendor.

Other data on taxes, crime, assessment, and other factors also change frequently. Your in-house data sources may also incur continuing expenses for collection, input, or reformatting.

Software upgrades are often released yearly with price tags that can run into hundreds of dollars. Additional programs may be required for new users, especially if the system is expanded to a network. Customer support may also require annual fees.

Equipment expenses to cover the cost of paper, print cartridges, hardware maintenance, and upgrades can add hundreds of dollars annually to your GIS bill.

Other expenses for training, salaries, office space, etc., can add up, but they are sometimes not directly factored into the firm's cost analyses.

Of course, you may not require most of the data sources noted. Or your company may already subscribe to the data sources you need and have all the equipment you will require. Depending on your GIS usage, software and hardware upgrades may be infrequent. This can help keep your recurring costs down.

Until the recent declines in software and data costs, companies and individuals in small or medium-sized real estate businesses were cautious about adopting GIS. Now that minimal start-up costs can be less than $1,000, interest in GIS and mapping is increasing. However, users should remember that recurring costs can exceed start-up costs within a relatively short period and should be carefully examined.

Cost-Benefit Analysis

GIS is a natural tool for larger organizations that have complex information systems and use numerous maps. For smaller firms, GIS requires technical expertise that may not be present or may be difficult to support.

Implementing a sophisticated GIS takes a significant period of time and, therefore, must frequently be justified in phases. Analyzing cost-benefits by phases is more practical in larger entities than in smaller businesses, which may be focused on day-to-day operations.

Ironically, smaller firms that implement GIS often enjoy the highest immediate paybacks, starting with mapping or incorporating GIS into integrated applications. Though small firms often lack the means to evaluate the cost-benefits of GIS properly, this will change as simpler, more customized programs become available.

Vendor Support

Customer Service

Customer support is often overlooked when hardware, software, or databases are acquired. The services that vendors provide vary considerably, as do their agreements, policies, and financial positions. For example, some vendors do not provide weekend technical support; such support may be unimportant to some but a necessity for others. The more complex the program and the less proficient the prospective user, the more important customer support becomes. The ultimate goal of good customer service is a satisfied customer. This can be accomplished with standard procedures or with some newer service innovations.

Technical Assistance

Occasionally, free technical service is provided for a limited period; thereafter, one must pay for support. The more sophisticated the GIS software, the more complicated the fee structure tends to be.

Although technical support is usually thought of as telephone assistance, it has evolved to include the other measures of support described below.

Telephone support. If you have problems with your hardware, software, or databases, you want them resolved quickly. Technical support can help but its effectiveness depends on:

- The hours and days of operation
- Toll-free long distance or 900 number access
- The knowledge level of the support staff
- The ability to get through and get your question answered

Making a phone call or two to technical support before purchasing a product can help determine how effective the technical support service will be.

Fax-back support. To handle routine questions, some vendors allow you to request documents by touch-tone phone 24 hours per day; the documents requested will be automatically faxed to you.[3]

3. Angela Hickman and Jodi L. Wynblatt, "The PC Magazine 150," *PC Magazine* (July 1994): 287.

Automatic voice support. Like fax-back support, this service provides recorded voice answers to common questions. It can be accessed toll-free and is also available 24 hours per day.

Internet. With rapid changes occurring with software, hardware, and databases, the ability to easily update data and upgrade software and hardware features can be essential. Web sites and file transfer protocol (FTP) sites that permit files to be downloaded and uploaded are becoming increasingly popular. These are similar but more convenient to use than a bulletin board service (BBS).

BBS. Before there were Web sites, BBSs allowed computer users to access needed software drivers, programs, or data quickly without waiting days to receive these items through the mail. A BBS requires a separate phone number, often a long-distance call, and cannot be accessed as inexpensively as a Web site.

Diagnostics. For some hardware, most often computers, a built-in diagnostics program can be used to identify problems. Some computer manufacturers include remote control software that allows technicians to directly diagnose problems on your computer.[4]

Warranty

Length. Many manufacturers provide a one-year, on-site warranty for computer hardware. Often the warranty can be extended for a fee to cover additional years. The total warranty for parts and labor is usually three years, but for certain parts it may be as long as five years. Traditionally, software support has covered the lifetime of the product, but it may soon be limited or provided on a fee basis.

Money-back guarantee. Although many vendors offer a 30-day guarantee, the length and meaning of the guarantee varies. The guarantee may apply only to product performance and not to customer satisfaction. If you are not entirely confident about a product, be sure it can be returned for any reason. A restocking fee may be required. Normally a product must be returned in its original box with shipping material; some vendors enforce this policy rigidly.

Return policies. Unfortunately, products can fail. Some companies will ship a replacement immediately, often by putting a temporary charge on your credit card; others require return shipment first. For hardware, many firms will only repair a product, while others will provide a new unit. The responsibility of paying shipping costs also varies.

On-Site Service Policies and Providers

Several vendors have excellent on-site service policies for hardware that eliminate the need to return a product. The timing and the geographical area covered depends on the particular provider, be it the vendor or a third-party repair service. On-site service may not be available, or it may automatically be included for one year. Sometimes it can be extended for another year, or even a few years, for an additional fee.

Repairs and reliability. Statistics indicating the frequency of repair or problems commonly associated with different hardware products may be available. Discussions with service providers and users can also provide insight into the reliability of hardware and software.

Delivery service. The savings offered by mail order firms often justify a purchase. Of course, if you order by telephone several days will elapse before the product reaches you, and shipping costs are extra. Delivery can be slow and expensive, especially for heavier hardware products. If the product is found to be defective shortly after purchase, it may need to be shipped back to the company, requiring much more time than if it were purchased and serviced locally.

Market Position and Stability

One issue of increasing importance is the likelihood that a vendor will continue to exist and provide adequate support. However, the size of a company does not ensure its financial stability. If a large

4. Rachel Derby Teitler and Carol Venezia, "Service and Reliability," *PC Magazine* (July 1994): 261.

purchase or commitment is being considered, a review of the firm's financial condition is advisable. Even if the vendor is a private firm, financial data can be procured from credit companies.

Another indication that can provide some assurance of a vendor's stability is its share of the market for a particular product or the total number of users. Without widespread acceptance of a product, especially a software package, it is difficult to find trained people to operate it. Hardware that is not widely used may not be readily supported by your programs.

Nevertheless, because GIS is an emerging field, a new product and company may be able to provide certain specialized capabilities that no other vendor appears likely to deliver in the near future. If this is the case, your specific needs may outweigh the vendor's market position and financial stability.

Agreements, Policies, and Fees

GIS software and data, especially customized versions, are usually purchased subject to agreements that can be severe and restrictive. For example, one vendor supplying street files expressly copyrights the maps created using its files and restricts the size of the map; printing maps 11-in. x 17-in. or larger would be a violation of the agreement.

Furthermore, software products are normally limited to one user and one computer. Network versions of software usually exist but allow only the stated number of users. For large installations, an unlimited site use license can sometimes be obtained. Also, certain programs still use copy protection schemes; if such a program no longer works on a computer, the copyright protection agreement may prohibit the reinstallation of that program on another computer.

Most programs have annual upgrades. If an upgrade is released shortly after a program is purchased, you may not need to pay for it. A few companies have annual service agreements covering technical support, manuals, training, and the like. A rider to this agreement can stipulate additional costs and fees. Depending on the vendor and the number of users, these annual service agreements may be negotiable. Service agreements may have a provision that automatically extends the agreement if notification is not received by a certain period prior to expiration. Failing to cancel a service agreement can be a costly mistake.

Staffing and Assistance

A broad-based application of GIS requires significant staffing and assistance, provided internally or externally. Mapping programs are usually more straightforward and require little or no assistance.

Fee Assistance

Assistance for specialized GIS needs can be obtained during implementation. A wide variety of consultants can provide special services to suit the particular demands of clients as well as general assistance. One might also turn to value-added resellers (VARs) or system integrators (SIs) for support.

Consultants. Typically, one consultant is not hired to do everything; most consultants specialize. One may have expertise in assessing your needs and identifying the type of GIS and data to acquire but lack the technical proficiency to install and test the system. New computer equipment or a network might demand a consultant with this expertise. Yet another consultant could customize the system with GIS scripts. The GIS software vendor might provide basic training for the system and supply ongoing support.

Compensation is normally charged on an hourly basis, but fixed-fee services and retainers can also be arranged.

Value-added resellers. VARs typically sell generic GIS software from the leading vendors as well as their own GIS software customized for real estate professionals. The applications of GIS are so varied and numerous that most GIS software companies rely on VARs to market their product to specialized market segments. This allows the GIS software company to expand quickly into a variety of markets while avoiding the mushrooming of its staff. A wide gamut of real estate services exist and each firm has its own particular needs, thereby creating a demand for specialized programs that can be provided by VARs.

Today there are relatively few VARs specializing solely in GIS and real estate. Their numbers are expected to grow as more real estate concerns find that their system needs can best be met through this type of service.

System integrators. Some large system integrators have expertise with every aspect of GIS systems: assessing needs; pulling together in-house data; recommending, selecting, and acquiring software and equipment; customizing software; completing installations; and providing for support and future growth of the system. Using such a firm to provide a turn-key operation is one of the easiest ways to install a large-scale GIS system, but it may also be one of the most expensive. It may be possible to assemble a knowledgeable team at more competitive rates by using a combination of consultants and a value-added reseller, but doing so can be quite a hassle.

Staffing Decisions

The inclination of many firms is to implement new systems and procedures entirely on their own. If an existing employee is eminently qualified to coordinate GIS, a firm is probably well able to take on the challenge. When a qualified individual or team is not available, staffing for GIS can be accomplished in one of two ways.

The first option is to train an employee who can then assume responsibility for the GIS system. This may require extensive instruction, especially if the employee is not well-versed in real estate or computers. If the designated individual or team does not have the proper background and training, he or she may be unable to implement GIS successfully.

If training existing staff is not feasible, one or more individuals who have experience and education in GIS could be hired. This can be expensive and the individual might require additional real estate training. Organizational issues relating to reporting procedures, supervision, and personal interaction may also be encountered as with any new hire.

Education and Training

In this discussion, a distinction is made between education and training. The former involves formal instruction at a school of higher learning. Training is job-specific and accomplished over shorter intervals. Three types of training—self-learning, non-fee training, and fee training—are discussed.

Education

Formal. Approximately 300 colleges and universities offer courses in Geographic Information Systems, and some have an entire accredited curriculum. Classes that concentrate on one particular GIS program are also available. Although GIS courses often provide general industry overviews only, some classes are beginning to cover applications specific to real estate.

General osmosis. In a 1994 survey, "GIS Strategies" determined that the greatest impediment to GIS expansion is ignorance about GIS and its uses.[5] The educational process is multi-dimensional and may involve business subordinates, peers, GIS marketing people, promotional literature, articles, etc. As more real estate-oriented establishments utilize mapping and GIS, more individuals will be aware of and educated about these tools. Fearing a competitive disadvantage, companies will place greater emphasis on GIS and mapping education.

Training

The most common method of learning GIS and mapping has been through on-the-job experience. Software documentation, quick-start and general reference manuals, on-line tutorials, and other forms of help are very useful when starting up a GIS. Researching the subject of GIS has become easier because more articles and texts (such as this book) have been published. Another source of information not to be overlooked are user groups and acquaintances who can provide support and ideas. Networking at meetings, lunches, and business functions such as real estate and GIS conferences also provides useful information.

5. Quoted from "GIS Strategies" in *Business Geographics* (July/August 1994): 18.

In-house training. Once an individual in your firm is trained in GIS, he or she can help train others. The in-house GIS training expert can be someone from information services, another department, or your own user group. A company might decide to hire someone new with the right education and/or job experience to be the firm's GIS expert. Agreements provided by some GIS vendors offer training at no additional cost.

Fee training. Organizations specializing in GIS, such as GIS World Inc. and certain software vendors, offer special GIS classes several times a year. Various real estate organizations are hosting seminars and workshops that address GIS as well. Arrangements can also be made with consultants and GIS experts to provide training sessions at their location or yours.

Implementation Alternatives

There is no single way to implement a GIS. According to many GIS articles, implementation seems to involve a high-end GIS and follow a top-down pattern. This has been the typical pattern when GIS is implemented at large corporations and government agencies. (See Figure 7.2.)

However, for real estate businesses a more limited implementation strategy may be applicable. A firm may begin with a stand-alone system or work group in one department and expand into other areas of the firm such as marketing, research, asset management, and appraisal.

A third type of implementation strategy is a bottom-up or grassroots approach using solution-specific mapping; this type of system may be adequate by itself or provide a platform to expand into GIS later. A small residential brokerage firm, for example, would most likely enter this new frontier through mapping.

Fourth, a GIS or its components can be integrated into an existing information system. Instead of GIS being the end-all solution, a mapping portion could be linked to a wide-scale functional aspect of a real estate business such as the accounting system.

Thus, GIS implementation does not need to include everything from the beginning. The system often needs to grow and progressive implementation might evolve in any number of ways. GIS can also be incorporated as part of a broader system or a specific application.

ORGANIZATIONAL ISSUES

The success or failure of a GIS installation is largely determined before the company leaves the starting blocks. Unless certain organizational issues are addressed beforehand, implementation can create endless problems. Organizational issues may be inconsequential if you have a sole proprietorship or if you control the final decisions in your firm. On the other hand, circumstances are far different if you work for someone else in a private company, a large corporation, a government entity, or any organization that requires decisions to be approved. A number of considerations may affect the GIS or mapping system you are planning to acquire or expand. Support for GIS or even simple mapping may depend on supervisory assistance, upper management support, departmental cooperation, the capability of your staff, and financial availability.

Even if you have upper management support, departmental cooperation is imperative if you need information located outside of your domain. GIS may require the assimilation of information from groups within your firm, and this process might threaten or encroach upon some individuals' responsibilities and power. If resistance develops, battlelines may be drawn. Some preliminary reconnaissance can help you identify hostilities and form alliances.

GIS can also create problems if it uncovers inadequacies in the firm's computer and information systems. Should this come as a surprise to top management, you and the GIS system might be perceived as the problematic factor. In such situations it is sometimes useful to have outside consultants deliver the bad news to ensure that the blame falls on past procedures, rather than you or your staff.

Political ramifications are often one compelling reason to begin GIS on a small scale. A limited project will not create undue alarm or concern among other departments and costs can be con-

Figure 7.2
Implementing GIS

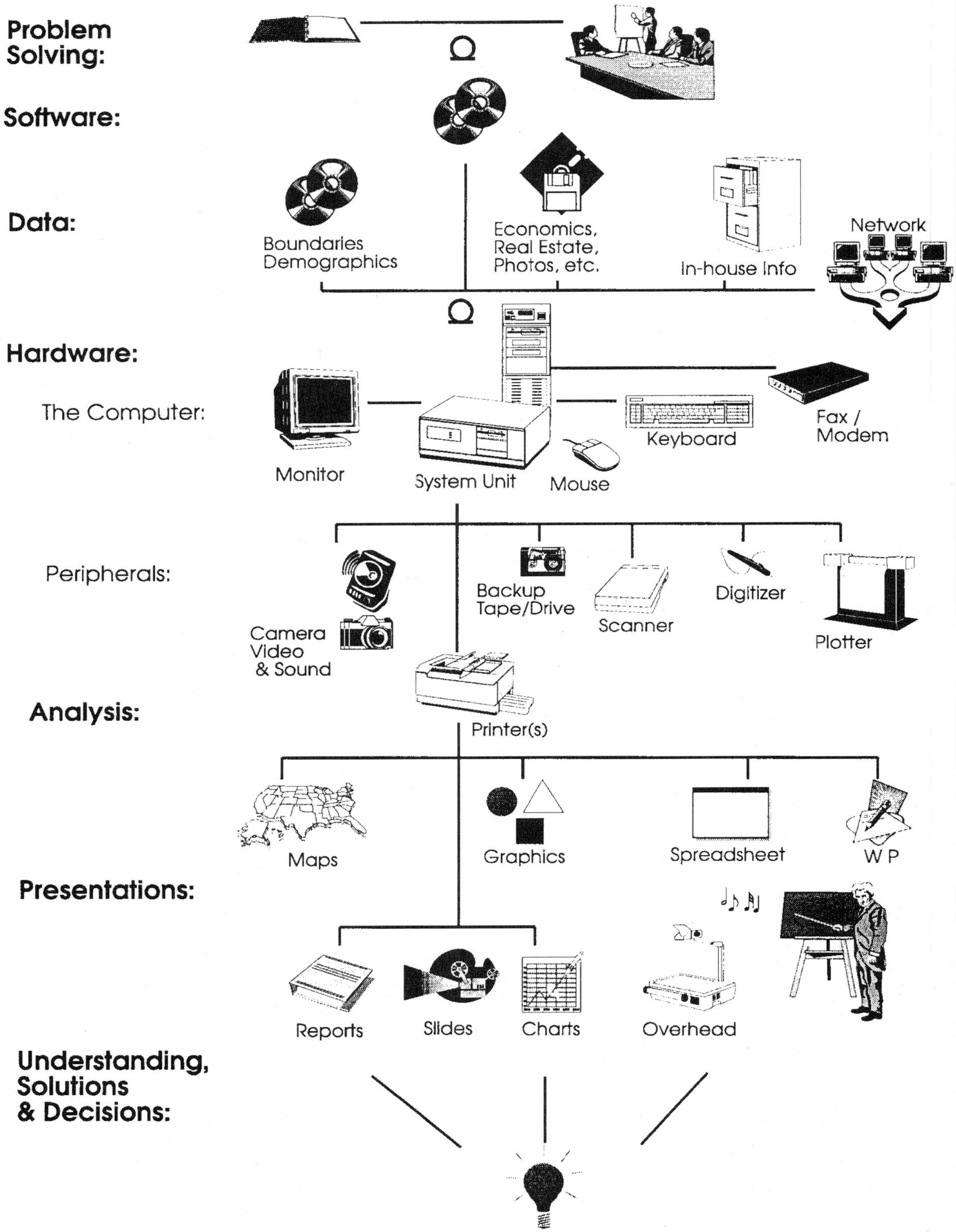

trolled. You or your staff can become more experienced and attain some short-term goals before long-term goals are established.

Preimplementation Studies

Before implementing GIS or mapping, clearly defined, realistic expectations should be developed. A bottom-up approach using a solution-specific mapping program or a limited- scale GIS requires few up-front considerations. However, much forethought must be given to any project in which GIS is spread across multiple departments and offices nationwide or worldwide. For large-scale installations, these studies might be needed:

- General needs assessment
- Function requirement analysis
- Cost-benefit analysis
- Recommended system configuration
- Strategic plan[6]

Answering Questions

When implementing or expanding a GIS system, be prepared to explain how and answer the five W questions—where, what, when, why, and who. You may have to confront a barrage of questions about the firm's goals and objectives and how GIS can help accomplish them. You may be asked

- What do we gain by doing this?
- When is it appropriate to begin and how long will it take?
- Where should GIS be implemented first? second? third?
- Why is it important or necessary to do this?
- How should we do this?
- Who should prepare a needs assessment, supervise the entire process, implement the system, and maintain it?
- Can it be done with existing personnel or must we hire others with certain qualifications?
- How should outside assistance be used?
- How much will it cost?

After procuring the necessary studies and identifying the specific uses of GIS, follow an agreed-upon plan. Using a project plan, with Gantt decision charts or similar planning aids, is more important for larger assignments. Some major issues that may be addressed in an implementation plan are described below.

Determination of Specific Goals and Objectives

- General coordination and time frame
  - Decide on a chief manager/coordinator
  - Identify in-house staffing
  - Identify outside assistance
  - Decide on each person's responsibilities and time frame for specific tasks
  - Review budget
- Specific tasks
  - Accumulate in-house data
  - Create/modify databases
  - Accumulate outside databases
  - Select software, including any add-on modules

6. Lynda L. Wiggins and Steven P. French, "GIS: Assessing Your Needs and Choosing a System," *American Planning Association PAS Report* No. 433 (Chicago: American Planning Association, August 1991): 16-17.

Select boundary files

Select hardware

New or expanded computer system

New peripherals or access to peripherals

Procure and set up hardware (and network connection if needed)

Procure and install software and data

Test compatibility of all hardware, software, and data

User orientation and training

Set up standard maps, presentations

- Prepare schedule—by phases if necessary

The implementation of a GIS will have some unforeseen consequences. These consequences can often be minimized by implementing the GIS more slowly, by department, by work group, or by individual office. The capabilities of the GIS can also be used in phases. For example, informational and marketing needs can be satisfied by GIS first, with administrative and management needs to follow. Especially with enterprise solutions from GIS, it is tempting to run before you walk, but it is not very practical.

FINAL COMMENTS

Until recently, GIS and mapping have had major limitations due to difficult-to-use software, expensive hardware, and limited data sources. The graphical user interface of new Windows-based mapping software makes it easier to use than with the old DOS interface. More powerful desktop computers, available at reasonable prices, provide performance levels that were previously only possible on mainframes, mini-computers, and high-end file servers. Desktop computer capabilities will dramatically improve as will the quality and performance of other equipment, including storage devices, scanners, digitizers, printers, and monitors. Prices have been spiraling downward for just about all GIS components including software, digital files, and attribute data. These trends are expected to continue.

While impediments to the adoption of GIS are crumbling, several hurdles remain. First, few real estate professionals are aware of GIS and its potential. GIS and computerized mapping are not yet common terms in real estate or many other business areas. This is changing with the introduction of mapping by major software behemoths like Microsoft. Mapping and GIS have too much potential to be ignored forever.

Second, specialized applications are needed to make the benefits of GIS more readily apparent. This will require the development of easy-to-use analytical techniques and algorithms that can be a standard part of special applications.

Third, prohibitive price tags on GIS software and data have kept potential customers away. Such products must be offered at reasonable costs.

Fourth, GIS demands substantial processing power, which has not been available at affordable prices until recently. Still more processing power is needed as is better hardware.

Fifth, for GIS to be widely adopted, demand must come from a broader market, especially from the bottom up, and link up to higher-end systems.

GIS and mapping are expected to be more seamlessly integrated into executive information systems (EIS) and management information systems (MIS). This capability may be developed by a system integrator, a value-added reseller, a consultant, or even an in-house user. However, at the present time the costs to deliver integrated systems to real estate are expected to be considerable. Since real estate encompasses many small companies, these all-in-one management systems, which include GIS and mapping, will probably be too costly for most real estate firms for a decade or more. Hence, a more bottom-up approach to GIS, starting with the use of mapping products or low-end GIS systems, will be the most logical path for most real estate firms for now.

chapter 8

Real Estate Information Technology: The Future Is Today

by Grant Ian Thrall, PhD

This final chapter of the text explores future uses of geographic information systems (GIS) software within the real estate industry and discusses the world of real estate information technology in general. Real estate, like the rest of society, cannot avoid participation in the new world of information technology. GIS is one important component of that new world and the World Wide Web is another. Together, GIS and the Web merge into a seamless technology that form the core components of real estate information technology.

In preparing this chapter, I spoke with many software development entrepreneurs developing products for GIS and the Web. Products under development that will soon be introduced to the marketplace are outlined and the overall message is clear: new developments are unfolding so rapidly that the future is today.

The future has arrived in part because of the manner in which ideas and technology are diffused across the landscape.[1] Often, a technology is introduced first in a single market such as San Francisco. As a result the potential technology adopters in other cities may not hear of the product and its growing popularity. The product may be introduced much later in, say, Cleveland. Eager technology adopters in secondary markets such as Cleveland can learn about their own future by observing what is happening now in other markets. So for Clevelanders, and perhaps for you, the future that is unfolding elsewhere can be experienced now in a form of time travel.

The way individuals respond to the new information age varies widely. Some will view the future described here as threatening. Others will see only opportunity and challenge. Our civilization has relied on advances in technology to increase efficiency. Thus we have developed the culture, standard of living, and general quality of life we enjoy today.[2] When GIS and the Web are adopted in the workplace, real estate agents, builders, lenders, appraisers, market analysts, and consumers will all benefit. The adoption of real estate information technology, however, comes at a price. The

Grant Thrall is responsible for the opinions expressed here and for any errors or omissions. Dr. Susan Elshaw Thrall contributed constructive and in-depth comments at all stages of manuscript development. Dr. A. Marks contributed to the conceptualization and early drafts of this material.

1. Richard Morrill, Gary Gaile, and Grant Thrall, *Spatial Diffusion*, Scientific Geography series, vol. 10 (Newbury Park, CA: Sage Publications, 1987).
2. Ibid.

market will witness a sizable decrease in the number of real estate agents, appraisers, and analysts needed to meet the requirements of the industry.

Spectacular real estate information technology is on the horizon and will usher in a more productive, more technological real estate industry. One can already see the impact of GIS in the real estate industry, where location is critical to every decision. In addition, the Web has made an impact,[3] creating a smaller world with more perfect information for the agent, seller, and prospective buyer. Using GIS we can measure the impact of location on each decision. With the Web, our own location need not place us at a disadvantage in obtaining information. For these and other reasons, GIS and the Web will become increasingly important in all aspects of real estate decision making.

RESIDENTIAL REAL ESTATE

Residential Brokerage

When it comes to the use of GIS in the buying and selling of homes, what will the future look like? Imagine that you are a potential home buyer. A real estate agent with a color notebook computer in hand comes to your home, or to your hotel room if you are in another city, and connects into the Web and the relevant Multiple Listing Service (MLS) Web site via a cellular phone. From the Web site the agent can access the traditional MLS database as well as many other computerized housing databases. These databases may compete with the MLS or represent an improvement on existing MLS databases. In any case, the databases will include listings of all newly built homes and all homes for sale by owner, and they will be available to help you, the home buyer, select your next home.

You begin your home search by answering questions on a form at the Web site identifying the style of home you want (e.g., ranch, Cape Cod), the features you prefer (e.g., number of bedrooms, number of bathrooms), and the amenities you desire (e.g., basement, fireplace, etc.). You can fill in the Web form from anywhere, but geography and geographic relationships (location) are perhaps the most significant determinants in the home buying decision. Thus, geographic relationships must be included as part of the computer search. "Location, location, location" is of paramount concern in real estate, and the Web will not change that anytime soon. GIS just makes location measurable.

As a potential buyer you answer a series of location-related questions on the Web form, including where you will be working and what distance or length of time you are willing to commute to work. You are also questioned about your preferred proximity to shopping and houses of worship and the desirability of a particular school district, city, town, or neighborhood. The questions also ask about preferred neighborhood amenities such as the availability of a community pool, the mix of single-family and multifamily homes, and demographic factors such as the percentage of households with children and their median ages.

Let us assume that you are connecting to the Web from your present home and will be relocating to a new city that you are not familiar with. You do not know the new metropolitan region or the surrounding towns and neighborhoods. Before the Web and GIS, you might have elected to rent a home for a year until you became familiar with the community or let your real estate agent steer you to a neighborhood that he or she thought you would like. This approach may have worked in the past, but it can be costly, time-consuming, and stressful. It is altogether much less efficient than moving directly into a home that you will enjoy.

Real estate information technology has a better solution. With the help of technology, you can take a computer walking tour of the communities that meet your desired criteria. The tour will include a map, both still and moving video images of key places of interest within these communities, and a soundtrack.

To narrow down your choice of possible new homes further, you will answer a series of financial questions designed to determine how much house you can afford. Your responses will be

3. Mauricio Rodriguez, J. Lipscomb, and W. Yancey, "Under All Is The Net," *Journal of Real Estate Literature* (Winter 1997); also published on the World Wide Web at http://zeta.is.tcu.edu/~mrodriguez/underall.htm.

immediately transmitted via the Web to a credit bureau, where a credit report will be prepared and sent directly to the mortgage lender of your choice. The information is later used in the mortgage lending process.

Once you have searched the Web database to find those homes that best satisfy your preferences and pocket book, you will be shown a map with the locations of the homes selected. You can then access the Web to obtain more detailed information on the targeted homes. Specifically, using the combination of the Web and GIS-based software, you are presented with pictures of the exterior of the house, the neighborhood, the street, the front and back of the property, a plat map, and a floor plan of the interior with complete room dimensions. If you are interested, you can take a complete electronic walk through the interior of the houses.

The information you have received from the computer tour is at least as detailed as what you could have learned by visiting the site—and this tour can be taken without leaving your present home, which may be thousands of miles away. Why bother with the computer tour? To save time and increase productivity. In the time it would take to visit one home, you could have previewed five, ten, or more homes by computer. Both you and your real estate agent have used your time more efficiently. In one or two hours, you can visit more houses in a much more relaxed environment than you would see in a week of driving around. And, when you do visit the city you are moving to, the search you have already performed will make your visit much more productive and much less stressful. If you conduct your final home search on your own, you can include your partner in the selection by placing images of the home chosen on the Web for him or her to view. With this type of information technology, a purchase decision could be made anywhere in the world in one or two days.

This approach is also much more efficient for the home seller. The home will be shown to more prospective buyers, but the home seller will be visited only by those buyers who are qualified and very seriously interested.

The apartment rental market will be affected by GIS and the Web in the same manner. A prospective renter can perform an on-line walk-through of the various apartments and unit floor plans within the building. Such a system is already being used successfully to advertise short-term rental accommodations at ski resorts in Vail, Colorado (see http://www.afn.org/~links/vail).

For some real estate firms and home buyers, the future described here is available today. Several firms are making such systems, or parts of those systems, available to individual real estate firms or local boards of real estate agents in selected areas. The impact of this change will be an increase in efficiency of real estate agents and a reduction in the number of agents required to sell the same number of homes.

Mortgage Lending

How will the integrated use of the Web and GIS change the process of establishing credit-worthiness and obtaining a mortgage? While you, the home buyer, are sitting with the real estate agent enjoying your computerized house tour, your detailed financial information is collected and transmitted via the Web to the mortgage lender of your choice. Your real estate agent can tie into various Web pages advertising the mortgage lenders doing business in the market area where your new home is located; these Web pages will also display the many loan options available. Because your agent is connected on-line with the lending institutions, the information is current. As information on rates and terms changes, the lending institutions immediately change what has been posted on the Web.

You select one or more possible lenders and, by the time you have completed your home-viewing session, the lender has reviewed your information and received your credit report. Assuming your credit is good, the lender can then issue a loan commitment letter back to you via Web e-mail. You are now ready to talk to the sellers. Your negotiating position as a buyer is considerably enhanced because you have already been qualified for a loan.

GIS has several other uses within the banking and mortgage lending industry. One more application in real estate lending is in the area of compliance, specifically with respect to the Community Reinvestment Act (CRA) and the nondiscriminatory granting of mortgage credit. One of the best ways for lenders to demonstrate compliance with the CRA is through mapping.[4]

First, demographic and census data are used to identify the low- and moderate-income census tracts within a lender's market area. Overlay maps can then be developed to show where mortgage applications originated and where mortgage loans were approved and/or denied. The maps will clearly show whether the lender has received mortgage applications from the residents of the low- and moderate-income census tracts within the market area and whether the lender is approving these mortgage loans. This type of mapping can also be used to study potential redlining and to show compliance with the Home Mortgage Disclosure Act (HMDA) requirements.

Real estate information technology software will include modules for obtaining credit reports and loan commitments. Computer software firms such as mine, Spatial Decisions & Analysis Inc., have already developed GIS software for CRA and HMDA compliance. GIS is already being used by many financial institutions for mortgage lending and the granting of credit. GIS will reduce the time and cost of processing a loan application and increase the efficiency of mortgage lending agents, which will reduce the number of agents required to process the same number of loans. Financial institutions that use Web and GIS technology will have an advantage over those who do not, and those who do not adopt these tools may be forgotten.

Appraisal

In the future GIS and the Web will be used together to determine the value of a house. An appraiser who works for a lender will have access to a computer that is linked to the Web databases described above and to the courthouse records for all real estate transactions. The appraiser will be able to select more appropriate comparable properties and to develop improved estimates of value using the extensive information on each house in the database. The information will include not only the amenities found in each house but also pictures of each room in both the home being appraised and all sales comparables. Such detailed information has seldom been available to appraisers; when it has been available, the cost of assembling the information has usually outweighed the benefits to be received. Real estate information technology is making the cost of assembling such information insignificant, so that all transactions, not just those involving very expensive properties, can be thoroughly investigated.

The new availability of housing databases facilitates both more accurate estimates of value and the routine application of generally accepted, sophisticated statistical analysis, including a combination of econometric hedonic regressions and geographic spatial statistics. In these statistical models, each amenity of the house (e.g., number of bedrooms, fireplace(s), hardwood floors) can be measured as to its contribution to the overall value of the property. The result could be appraisals of unprecedented accuracy. Moreover, an appraisal prepared using a database could be completed in an hour or less and would include computer-generated forms, supported by all the necessary pictures, maps, and floor plans.

The real estate information technology software described earlier also includes the new GIS appraisal capability. Academically oriented appraisers and geographers (such as Brian J.L. Berry at the University of Texas and myself at the University of Florida) are presently working on the development of GIS-based hedonic regression models for the determination of residential property value. The application of GIS to residential property appraisal is happening now and will reduce transaction costs and lead to more timely, complete residential appraisals.

Housing Demand

Can GIS be used in forecasting the demand for housing by price and location? In the past analysts have looked at the historical average growth within an area, such as a county or submarket, and

4. Grant Ian Thrall, Judy Fandrich, and Susan Elshaw Thrall, "The Location Quotient: Descriptive Geography for the Community Reinvestment Act," *Geo Info Systems,* vol. 5, no. 6, 1995, 18-22, and Grant Ian Thrall and Paul Amos, "GIS Within Real Estate and Housing Mortgage Industries: Agendas for the Near Future," *Business and Research Uses of GIS in Housing Mortgage Finance.* Proceedings of the 1996 Fannie Mae Roundtable Conference. (Washington, D.C.: Fannie Mae, 1996).

attempted to forecast the future demand for housing in that area based on the number of housing sales and past absorption patterns. Since we recognize the importance of location and the impact of each property location on surrounding properties, forecasts would be more accurate if they included the geographic component along with the traditional time component used in forecasting models.

In other words, assume that we want to know the housing absorption rate for a specific neighborhood in the year 2000. The time-geographic forecast would take into consideration four factors:

1. The historical growth of the neighborhood, e.g., a census tract.
2. The historical growth of the neighborhoods surrounding the target neighborhood.
3. Limitations on future growth attributable to zoning and/or environmental constraints.
4. The expected population increase for the region (i.e., the county) surrounding the neighborhood. Geocoded parcel-specific data would be required to perform the GIS application.

Such an application is possible today for small-area forecasting. In 1993 my colleagues and I presented the cascade GIS diffusion model in an article in the *Journal of Real Estate Research.*[5] The cascade model was accurate for small-area forecasting with an error that generally did not exceed 3%.

COMMERCIAL PROPERTY

Office

What about the combined use of the Web and GIS to locate office space? Real estate agents have long used a multiple listing service to assist both home buyers and home sellers. Because each participating agent in the area has access to the MLS and information on each house listed for sale, the number of potential buyers who receive information about a particular house is increased as is the likelihood of a sale. Wouldn't it be useful to have on-line, GIS-based MLS for commercial real estate, like the one described in the residential brokerage section? Such an MLS could assist commercial real estate agents in leasing office space.

In fact, several such systems are already operational. A commercial real estate agent can sit at a computer terminal and search a CD-ROM for the office building that meets the client's requirements in terms of location, amount of square footage needed, desired rent level, and amenities (e.g., Class A space with an atrium lobby and underground parking). The information presently on the CD-ROM is expected to be available on the Web soon. The agent will search the Web spatial database and request a map display showing the location of each office building that meets the client's search criteria. By clicking on a particular building location, the agent can specify a still or moving video image of the exterior of the building. The program user also has the ability to display the floor plan of each floor along with the current tenants and the square footage they occupy.

With this real estate information technology, the user can undertake searches by location, which is critical in finding an optimal office site. Agents can search the entire market or a submarket in a few minutes and develop proposals for alternative sites that meet the client's criteria. Via the Web this information can be accessed from any part of the world for any part of the world. This GIS application increases an agent's efficiency, reduces the cost to the client, and enhances profitability for the building owner. In short, this technology unites the world into a global real estate market.

It is also possible to use GIS and Web databases to forecast the demand for office space by applying a cascade-like diffusion model.[6] The user looks at the square footage of office space developed in a relevant submarket to generate a forecast of the amount of office square footage required in that submarket at some future time. This type of forecasting is possible only with the use of a GIS and the introduction of easily accessible, location-specific data for each office building.

5. Grant Ian Thrall, Charles Sidman, Susan Thrall, and Timothy Fik, "The Cascade GIS Diffusion Model for Measuring Housing Absorption by Small Area with a Case Study of St. Lucie County, Florida," *Journal of Real Estate Research* (Summer 1993): 401-420.

6. Ibid.

Retail

Retail real estate markets differ significantly from office markets, and the number of retail sites in a market far surpasses the number of office buildings. Along with regional malls, neighborhood retail centers, and small shopping strips, freestanding outlets such as Blockbuster Home Entertainment stores and many other individual stores dot the landscape.[7] Identifying each of these locations and setting up a complete MLS for retail uses would be as difficult as trying to add all home offices into an office market database. However, in a few cities attempts are now underway to create a Web-accessible, GIS-based MLS for retail uses.

More interesting is the use of GIS by retailers to identify optimal site locations. With the availability of the TIGER/Line geographic database from the 1990 U.S. Census, most major retail store chains and their consultants now use GIS to identify potential markets at the macro level of analysis and optimal store locations within those markets at the micro level of analysis. They rely on a variety of tools and procedures, which were reviewed in a series of articles by my colleagues and myself in 1996.[8] Among these tools is the geographer's gravity and spatial interaction model, which is used to identify potential market areas and estimate expected market area penetration.[9]

In the past century geographers have developed a number of gravity-like models, which vary in their complexity and accuracy. At one end of the spectrum, the market models used by retail businesses can be as simple as identifying and mapping the homes of their customers. These GIS-based models are analogous to Applebaum's customer spotting model in traditional real estate analysis,[10] except that the map and pins are electronic and displayed on a computer monitor. At the other end of the spectrum are Huff spatial interaction models, which can mathematically calculate the probability that someone living at one location will shop for an item at another location. These more complex models can determine probabilities based on hypothetical retail locations; calculations are performed for each possible location identified on the landscape. The result is a potential probability surface. The location with the highest potential (highest expected revenue) would be included in a retail location evaluation.

Some gravity and spatial interaction models require highly sophisticated GIS algorithms that are not available in typical commercial GIS software packages.[11] The demand for these types of GIS tools by real estate and other industries is driving the GIS industry in two directions. This phenomenon is discussed in the last section of this chapter.

GIS is clearly useful for retail site selection analysis, but where are the other GIS applications in the retail real estate market? Imagine defining a market area for a retail center and then analyzing the demographics of the current neighborhood by census block. Such an analysis would include information on household composition with respect to size, age, income, and expenditure patterns, right down to the amount of money spent on dental floss. Now imagine that we can take these detailed expenditure data and convert them into indications of the demand for the various types of retail stores that could be within the center. Next we perform an analysis of the competition and develop a profile of the type of stores that would optimally locate in the center. We can now give the leasing manager a list of desired stores and the anticipated revenue per square foot that will be generated by each retail establishment. Before signing a lease, the leasing manager can easily

7. Grant Ian Thrall, Mark McClanahan, and Susan Elshaw Thrall, "Ninety Years of Urban Growth as Described with GIS: A Historic Geography," *Geo Info Systems,* vol. 5, no. 4 (April 1995): 20-27.
8. Grant Ian Thrall, Juan del Valle, Paul Amos, and Ken McGurn, "Measuring the Development Potential of an Historic Downtown," *Geo Info Systems,* vol. 6, no. 4, 1996, 44-49; Grant Ian Thrall and Juan del Valle, "Retail Location Analysis: Antecedents," *Geo Info Systems,* vol. 6, no. 6, 1996, 48-52; Grant Ian Thrall and Juan del Valle, "Marketing Geography Innovations of William Applebaum," *Geo Info Systems,* vol. 6, no. 8, 1996; Grant Ian Thrall and Juan del Valle "Calibrating An Applebaum Analog Market Area Model With Regression Analysis." *Geo Info Systems,* vol. 6, no. 11, 1996.
9. Kingsley Haynes and A. Stewart Fotheringham, *Gravity And Spatial Interaction Models,* Scientific Geography series, vol. 3. (Newbury Park, CA: Sage Publications, 1984).
10. Grant Ian Thrall and Juan del Valle, "Retail Location Analysis: Antecedents," *Geo Info Systems,* vol. 6, no. 6, 1996, 48-52; Grant Ian Thrall and Juan del Valle, "Marketing Geography Innovations of William Applebaum," *Geo Info Systems,* vol. 6, no. 8, 1996; Grant Ian Thrall and Juan del Valle, "Calibrating An Applebaum Analog Market Area Model With Regression Analysis." *Geo Info Systems,* vol. 6, no. 11, 1996.
11. Grant Ian Thrall, "Modular Component Programming: The Foundations of GIS Applications," *Geo Info Systems* (February 1996): 45-46; "SylvanMaps/OCX," *Geo Info Systems* (February 1996): 47; "Maps, Data & Mapplets: First Impressions Software Review," *Geo Info Systems* (February 1996): 48.

determine the tradeoff in overall center revenue that results when one store type is selected over another.

This type of GIS-based retail analysis is available today. Detailed expenditure data at the level of the census block can be obtained from several secondary data vendors and GIS-based decision models are presently being developed by entrepreneurs in the GIS industry. Such GIS applications will ultimately make retail site selection and the identification of the best mix of stores at a retail site more efficient and more profitable.

We have already noted that the cascade diffusion model can be used to forecast the demand for residential real estate and the required square footage of office space. The cascade diffusion model can also be used to forecast the demand for retail space at the submarket level, by factoring in the actual square footage developed over time at the parcel level.

Industrial

The number of residential sales transactions in most real estate markets greatly exceeds the number of industrial transactions. In any given submarket, there may be a small handful of industrial listings. Industrial relocation and expansion are not daily occurrences. In addition, the market area for industrial real estate may encompass one state or a group of states, while the market area for residential real estate may include only a small part of a metropolitan area. Information on industrial real estate is decidedly more complex but, with the combined use of the Web and GIS, professionals can bridge the information gap and obtain an understanding of large industrial real estate markets.

One firm in Atlanta, Georgia, monitors industrial real estate listings throughout the Southeast. Businesses interested in industrial property in the Southeast provide the firm with their criteria, such as the desired maximum distance from a 5,000-ft. airport runway and the maximum distance from an urban area with a population of more than 500,000. Businesses can also specify the characteristics they are looking for in the area, e.g., average unemployment rate, average wage rate, frequency of labor unrest, and their minimum requirements for the quality of local education, parks and recreation, and other amenities. Using appropriate GIS software and the database, which was developed and is maintained at great expense, the firm can almost instantly respond to business inquiries about available industrial real estate.

Local municipalities could identify industrial properties that are vacant or for sale on a Web page. Any commercial real estate brokerage could have its own Web search engine automatically retrieve and store changes in the industrial listings posted on the Web in its database. Use of the Web would make for a much more efficient industrial real estate market and assist in local economic development.

Several states have completed a statewide GIS to provide information to firms considering relocation to those states.[12] Using the GIS, state economic development officers can instantly respond to queries from manufacturers about the most appropriate location for their particular industrial activity, which gives the state an edge in attracting industry.

Real estate information technology will soon revolutionize industrial real estate transactions throughout the nation.

IMPLEMENTATION FACTORS

Geographic Data

How will the availability of geographic data change in the future? Most GIS users look forward to the day when they can perform address-specific, street-level geocoding. Unfortunately, even with improvements to the TIGER/Line geographic databases, not all new streets are available in the electronic versions of the U.S. street network. As a result, success rates for geocoding typically fall

12. Robert Pittman and Grant Thrall, "Using Geographic Information Systems In Economic Development," *Economic Development Review* (Fall 1991): 14-21 and "Improving Real Estate Market Research," *Real Estate Issues* (Spring/Summer 1992): 1-7.

in the 50% to 75% range. In fact, a careful reading of information from geographic data suppliers reveals that even vendors who claim to offer enhanced and improved geographic data have not significantly increased the number of new street segments.[13]

The real estate market wants street files that are updated monthly with near-perfect street-level geocoding. Several geographic data providers plan to make updates to street data files on a monthly basis, borrowing from the U.S. Postal Service routing tapes that are available on a monthly basis. There have been discussions about the U.S. Postal Service taking over responsibility for updating the Tiger/Line street databases since its workers travel the nation's streets on a daily basis. Postal personnel could and should carry global positioning system (GPS) devices as they travel their routes, updating the Tiger street database with new streets and address changes continuously.

Today, however, obtaining usable hit rates translates into geocoding to polygons such as the centroid of a census tract or a zip + 4 area.[14] Several geographic data vendors offer zip + 4 centroid files. In fact, most vendors create their centroids by geocoding a fictitious set of data to a street address based on their most current or enhanced TIGER/Line street line segments. They then look for a "hit" and the resulting coordinates serve as the coordinates for the zip + 4. Most vendors offer semiannual or quarterly updates.

It is possible to develop a mathematical procedure for identifying the centroid of a zip + 4 boundary using postal routes. Mathematical procedures that go beyond U.S. Census offerings of TIGER/Line street data or manually improved street line databases might offer a higher level of accuracy. Research based on quantitative geographers' algorithms is highly marketable and should soon become available. The availability of better, less expensive zip + 4 coordinates and, ultimately, the ability to geocode to street address will increase the accuracy, value, and productivity of the analyses performed with a GIS.

GIS Software

What does the future hold for GIS software products? Wouldn't it be valuable to have user-friendly GIS software that meets the user's exact needs, can be embedded in a given application, offers a complete range of GIS functionality, and could be used simply by pressing one or two buttons? This futuristic vision is the present state-of-the-art among software developers.[15]

In the fall of 1994, the release of Lotus 1-2-3 version 5 included a breakthrough for popular acceptance of geography software: built-in thematic mapping capability from the now-subsumed GIS software vendor Strategic Mapping Incorporated. This was followed shortly afterward by the release of Microsoft's new version of Excel, which included elementary functionality from the GIS software program MapInfo.[16] While neither version offered GIS functionality, their thematic mapping capabilities heightened public awareness that geography offers analytic utility and is more than a tool to play "name the capital." At about the same time, mainstream GIS software manufacturers announced new pricing for their Microsoft Windows GIS software—an attractive price range below $400. Thus, the future looks bright for converting those who presently rely on database management and spreadsheet software to the use of desktop GIS software.[17] An even more exciting development is the advancement of GIS component programming tools, called OCXs.[18]

These tools for software developers cut software development time dramatically. It is now financially feasible for mapping capability to be designed into a single-user custom software program; the development costs for such program can compete with the general, do-it-all GIS software of several years ago.

13. Grant Ian Thrall and Susan Elshaw Thrall, "Business GIS Data, Part Two: Highend Tiger/Line," *Geo Info Systems,* vol. 3, no. 9, 1993, 66-70 and "Business GIS Data, Part Three: ZIP Plus 4 Geocoding," *Geo Info Systems,* vol. 4, no. 1, 1994, 57-60.
14. Ibid. and Grant Ian Thrall, Juan del Valle, and Susan Elshaw Thrall, "Business GIS Data Part 6: When is ZIP+2 Geocoding Good Enough?" *Geo Info Systems,* vol. 5, no.11, 1995, 40-43.
15. Grant Ian Thrall, "Modular Component Programming: The Foundations of GIS Applications," *Geo Info Systems* (February 1996): 45-46 and "SylvanMaps/OCX," *Geo Info Systems,* February 1996, 47.
16. Reviewed in Grant Ian Thrall, "Maps, Data & Mapplets: First Impressions Software Review," *Geo Info Systems* (February 1996): 48.
17. Grant Ian Thrall, "New Generation of Mass Market GIS Software: A Commentary," *Geo Info Systems,* vol. 5, no. 9, 1995, 58-60.
18. _______, "Modular Component Programming: The Foundations of GIS Applications," *Geo Info Systems* (February 1996).

In the computer software market there is an inevitable movement toward lower costs and increased ease of use. With the OCX technology, GIS and thematic maps can now be embedded in a software application. The inevitable increase in demand for higher levels of GIS functionality—beyond simple thematic mapping—will lead to greater demand for geographic and demographic data.

Many GIS software manufacturers have adopted the philosophy that their GIS products can be used to solve the problems of any industry. Some manufacturers of GIS software allow for customization via their scripting language, thereby allowing their general-purpose GIS software products to be modified to appear industry-specific or even firm-specific. However, because most scripting languages are unique to the GIS software, most users are unable to modify their GIS to fit their needs. Moreover, most managers have neither the time nor the inclination to become GIS programmers or to acquire a professional geographer's level of spatial reasoning. For these reasons, many users are now contracting out the development of user-friendly, specially designed, GIS products.

A small group of companies (such as mine) specialize in developing user-specific and market-specific GIS applications. Examples include the home-viewing system described above, the cascade diffusion forecast model, CRA bank compliance, retail location strategy, and portfolio management. These products are not of the "one software program fits all" variety. Instead, by using rapid-development software tools and becoming industry- or even client-specific, a few firms are developing software with greater user-friendliness and user-applicability. The software can be written to include only those features and algorithms required to accomplish the specific tasks desired by the client and the software interface can be designed to reflect the general procedures followed by the firm. Instead of being a complex hurdle for the client, the GIS functionality can be totally transparent and automatic, with custom software that performs the desired task with a click of a button.

Real Estate Information Technology Personnel

Some operations can be made routine and hidden within the software Click a button and the analysis is performed. But what type of geographic analysis is appropriate for a particular business decision? Since the discipline of geography was established, many complex types of geographic analysis have been undertaken. However, few individuals have mastered the domain of quantitative geography and the use of GIS and the Web. This has created a market for consulting and custom software development firms. As more real estate decisions depend on sophisticated geographic analysis, the greatest benefits may go to whoever is able to assemble the Web/GIS war room first and those who are using the correct geographic algorithms for their business decisions.

Where will the geography gurus of the future come from? When I was consulting with a major accounting firm on real estate information technology, a senior partner complained of the lack of qualified personnel. He pointed to a wall of plaques identifying the various universities at which the firm had established endowed professorial chairs and said, "With the large investment my firm has made in higher education, you would expect that we would be able to hire properly trained personnel!" I responded with a question: "How many of those endowed chairs are designated for geographers?"

That is the dilemma facing the industry today. Real estate information technology is revolutionizing the industry, and businesses want to employ people with the appropriate training. However, university faculties have experienced a greater reduction in real income than any other occupational group, which has resulted in a shortage of faculty in high-demand fields such as real estate information technology

Consider the following true story. I recently graduated three students: one with a BA, one with an MA, and one with a PhD. All the students were very good. The student with the BA completed four years of college and had no on-the-job experience. The student with the MA completed six years of university and he too had no on-the-job experience. The student with the PhD completed 10 years of university education and had four years of practical on-the-job experience in the field of

GIS. The BA and MA graduates both received job offers from highly visible Fortune 500 firms. The PhD student received a job offer from a highly regarded, nationally recognized university. The salary offered to the BA student was 15% less than the salary offered the student with the PhD. The salary offered to the MA student was 20% more than the PhD student received. The question this situation poses is: Will the future be clouded by no one qualified to train, students graduating with questionable qualifications, and an underqualified work force with "Dilbert-like" management?

Universities do not have the funds to pay market-competitive faculty salaries or create new faculty positions in response to changing market demand. University budgets are not market-driven—they are politically driven. Public universities must compete with prisons for funding. Deans presiding over departments of geography or colleges of business may be inclined to dismiss evidence of the change in demand for appropriately trained geographers because they may be asked to increase salaries and faculty positions with funds they do not have. Coincidentally, the shortage of trained students and the shortage of university faculty qualified to provide training and advance the technology may bring a sigh of relief to those who have become fearful that the new real estate information technology will prove financially ruinous.

For those who want to see the future of real estate information technology fully realized, there is one solution—money. There must be an unprecedented investment in the creation of endowed chairs to promote scholarship in the field of geography.

CONCLUSIONS

What is the future for GIS in real estate? In real estate, perhaps more than in any other sector of the economy, the ability to measure the impact of location is critical. Only with the advent of GIS has it become possible for practitioners to measure the true impact of location. Information is required for the GIS to accomplish this task and the Web has become the necessary information provider. GIS software programs are becoming more user-friendly and less expensive. The quality of the available geographic data is improving and the future looks bright. The only obstacle in sight may be the limitations of an education system that has not allocated sufficient funds to train individuals to maintain and develop innovative GIS technology.

appendix 1

Additional GIS Sources

by Gilbert H. Castle III

GIS LITERATURE, ORGANIZATIONS, ETC.

Publishers of GIS Books and Periodicals

GIS World, Inc.
http://www.geoplace.com

Geo Info Systems
http://www.geoinfosystems.com

GIS Glossaries

GISLinx
http://www.gislinx.com/GISLinx/GIS_jargon.htm

Jefferson County (Colorado) GIS
http://ozma.jefferson.co.us/dpt/gis/glossary.htm

Association for Geographic Information (AGI)
http://www.geo.ed.ac.uk/agidict/welcome.html

British Columbia Ministry of Environment, Lands, and Parks
http://www.env.gov.bc.ca/gis/glosstxt.html

ESRI, Inc.
http://www.esri.com/base/users/glossary.html

GIS Professional Societies

Urban and Regional Information Systems Association (URISA)
http://www.urisa.org

AM/FM International
http://www.amfmintl.org

GeoBusiness Association
http://www.gba.org

Miscellaneous

Federal Geographic Data Committee — Clearinghouse
http://www.fgdc.gov/Clearinghouse/Clearinghouse.html

International Real Estate Directory
www.ired.com

Peter Pike's "Pikenet"
www.pikenet.com

GIS VENDORS

NOTE: The vendors are listed under one of the following four categories: Software, Digital Data, Attribute (Tabular) Data, or Services. Many vendors offer products and services in two or more of these four categories.

Software—Including Internet Tools

Autodesk Inc.
http://www.autodesk.com/gis

Caliper Corp.
http://www.caliper.com

Cypress Geo-Resources Inc.
http://www.cyp.com

Decisionmark Corp.
http://www.decisionmark.com

ERDAS Inc.
http://www.erdas.com

ESRI Inc.
http://www.esri.com

Genasys
http://www.genasys.com

GeoResearch Inc.
http://www.georesearch.com

GeoSystems Global Corp.
http://www.mapquest.com

The IDRISI Project
http://www.idrisi.clarkeu.edu

Insurance Services Office Inc.
http://www.iso.com

Intergraph Corp.
http://www.intergraph.com

MAPCOM Systems Inc.
http://www.mapcom.com

MapInfo Corp.
http://www.mapinfo.com

MatchWare Technologies Inc.
http://www.matchware.com

Northwood Geoscience Ltd.
http://www.northwoodgeo.com

Object/FX Corp.
http://www.objectfx.com

Qualitative Marketing Software Inc.
http://www.qmsoft.com

SAS Institute Inc.
http://www.sas.com

Scan/US Inc.
http://www.scanus.com

Sites USA
http://www.sitesusa.com

Tactician Corp.
http://www.tactician.com

Tetrad Computer Applications
http://www.tetrad.com

Total Systems Inc.
http://www.totalsystemsinc.com

Visio Corp.
http://www.visio.com

Digital Data—Including Vector, Raster, Imagery & Data Conversion

American Digital Cartography Inc.
http://www.adci.com

BLR Data
http://www.blrdata.com

DeLorme
http://www.delorme.com

Desktop Mapping Technologies
http://www.desktopmapping.com

Environmental Risk Information & Imaging Services
http://www.eriis.com

Geographic Data Technology Inc. (GDT)
http://www.geographic.com

GeoImage Technologies
http://www.geoimage.com

HJW & Associates
http://www.hjw.com

Image Scans Inc.
http://www.imagescans.com

ImageLinks Inc.
http://www.imagelinks.com

Landcare Aviation Inc.
http://www.landvue.com

LizardTech Inc.
http://www.lizardtech.com

Magellan Systems Corp.
http://www.magellangps.com

The MapFactory Inc.
http://www.mapfactory.com

On Target Mapping
http://www.otmapping.com

RMSI
http://www.rmsinet.com

Rolta International Inc.
http://www.rolta.com

Sedona GeoServices Inc.
http://www.sedonageo.com

Silvana Inc.
http://www.cogno.com/silvana

Space Imaging EOSAT
http://www.spaceimaging.com

SPOT Image Corp.
http://www.spot.com

TGS Technology
http://www.tgstech.com

Thomas Brothers Maps
http://www.thomas.com

3D Imaging LLC
http://www.3diterralogic.com

TRIFID Corp.
http://www.trifid.com

TruVue Corp.
http://www.truvuecorp.com

ValueCAD
http://www.valuecad.com

VARGIS LLC
http://www.vargis.com

Attribute (Tabular) Data

Applied Geographic Solutions Inc.
http://www.appliedgeographic.com

The Buxton Co.
http://www.buxtonco.com

CACI Marketing Systems
http://demographics.caci.com

Claritas Inc.
http://www.claritas.com

Compusearch Micromarketing Data & Systems
http://www.polk.ca

Easy Analytic Software Inc.
http://www.easidemographics.com

Experian (formerly TRW Redi)
http://www.experian.com

Geoscape International Inc.
http://www.geoscape.com

Market Statistics
http://www.marketstats.com

National Decision Systems
http://www.natdecsys.com

NPA Data Services Inc.
http://www.npadata.com

Online Data Services
http://www.onlinedatasvc.com

TGE Demographics
http://www.wessex.com/TGEDemographics

VISTA Information Solutions
http://www.vistainfo.com

Services—Including Consulting, Programming, & Systems Integration

(See also the Web sites of the authors of this book.)

ADR Inc.
http://www.adrinc.com

Applied Geographics, Inc.
http://www. appgeo.com

Berger & Co.
http://www.berger.com

Castillo Co. Inc.
http://www.castillo.com

Convergent Group
http://www.convergentgroup.com

Dakota Worldwide Corp.
http://www.dakotaww.com

Data Insight
http://www.datainsight.com

Dynamic Geographic Technologies Inc.
http://www.dgt-inc.com

Find
http://www.find-gis.com

Geo Insight International Inc.
http://www.geoin.com

Geodata Consultants Inc.
http://www.geodataconsult.com

Geodyssey Ltd.
http://www.geodyssey.com

Geographic Services Corp.
www.MapAgent.com

Geomatics International Inc.
http://www.geomatics.com

Integrated GPS Technologies Inc.
http://www.gpsgis.com

Komex International Ltd.
http://www.komex.com

Mapping Analytics
http://www.mapanalytic.com

Mapping Technologies Inc.
http://www.maptech.net

MapPix
http://www.mappix.com

MarketQuest
http://www.marketq.com

Matrix Research LLC
http://www.matrixr.com

Michael Baker Jr. Inc.
http://www.mbakercorp.com

MPSI Systems Inc.
http://www.mpsisys.com

Site Perfect Research Inc.
http://www.siteperfect.com

Spatial Insights Inc.
http://www.spatialinsights.com

appendix 2

Color copies of some of the maps shown in Chapters 2, 4, 5, and 6 are included in this section.

Figure 2.1

Town of Framingham, Mass

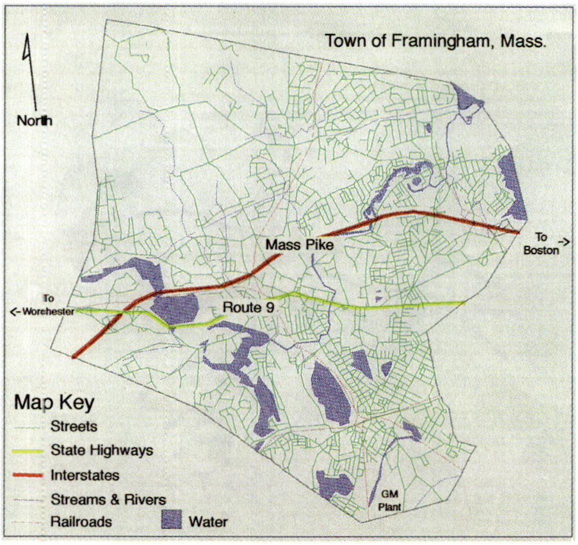

Commercial Property Annual Income/Sq. Ft.

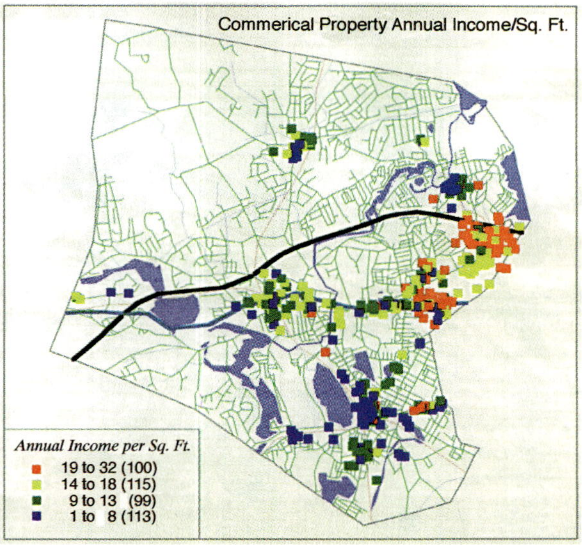

Figure 2.3

GIS Maps Tailored to Users

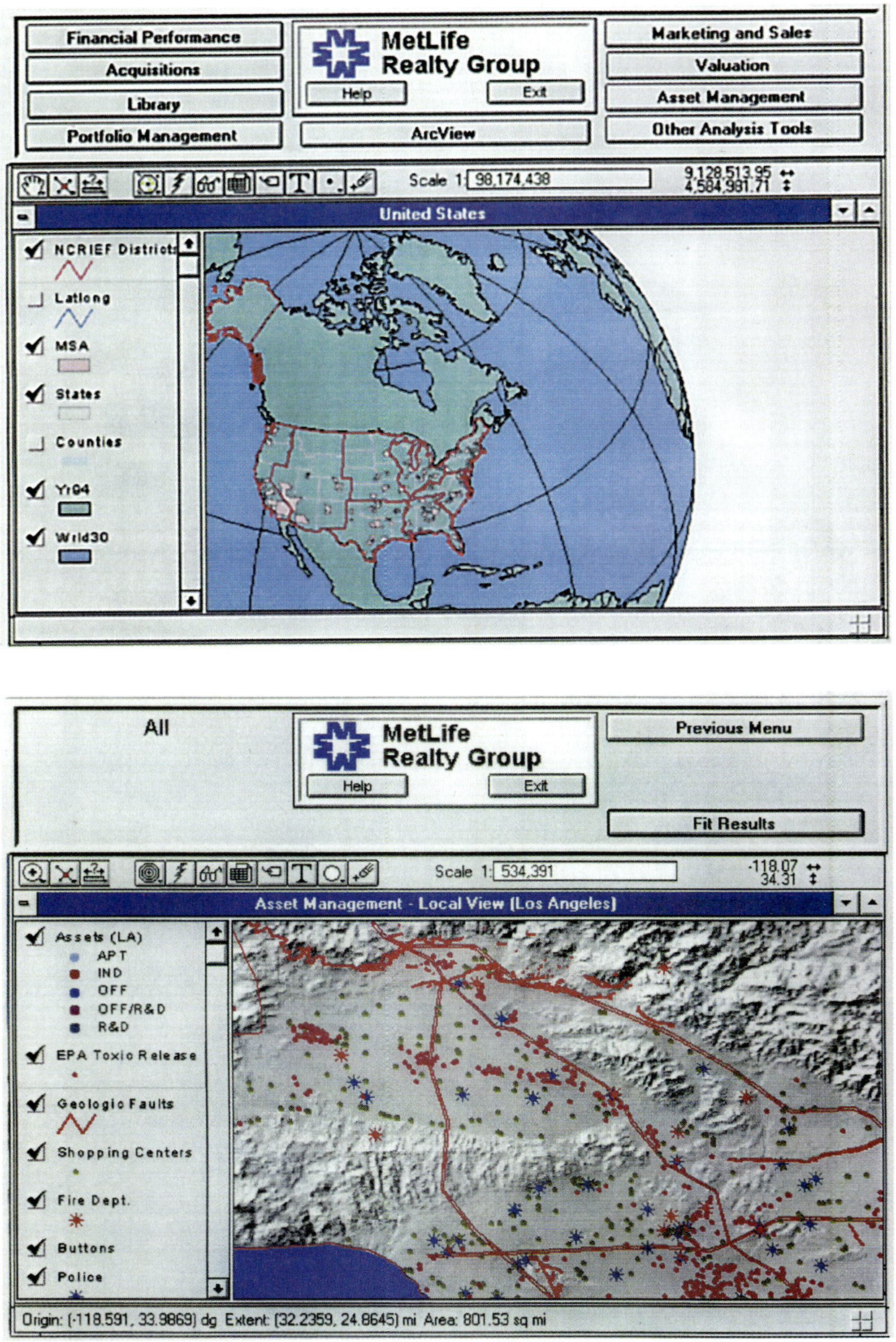

Figure 4.7
Bear Valley—Hydrant Location

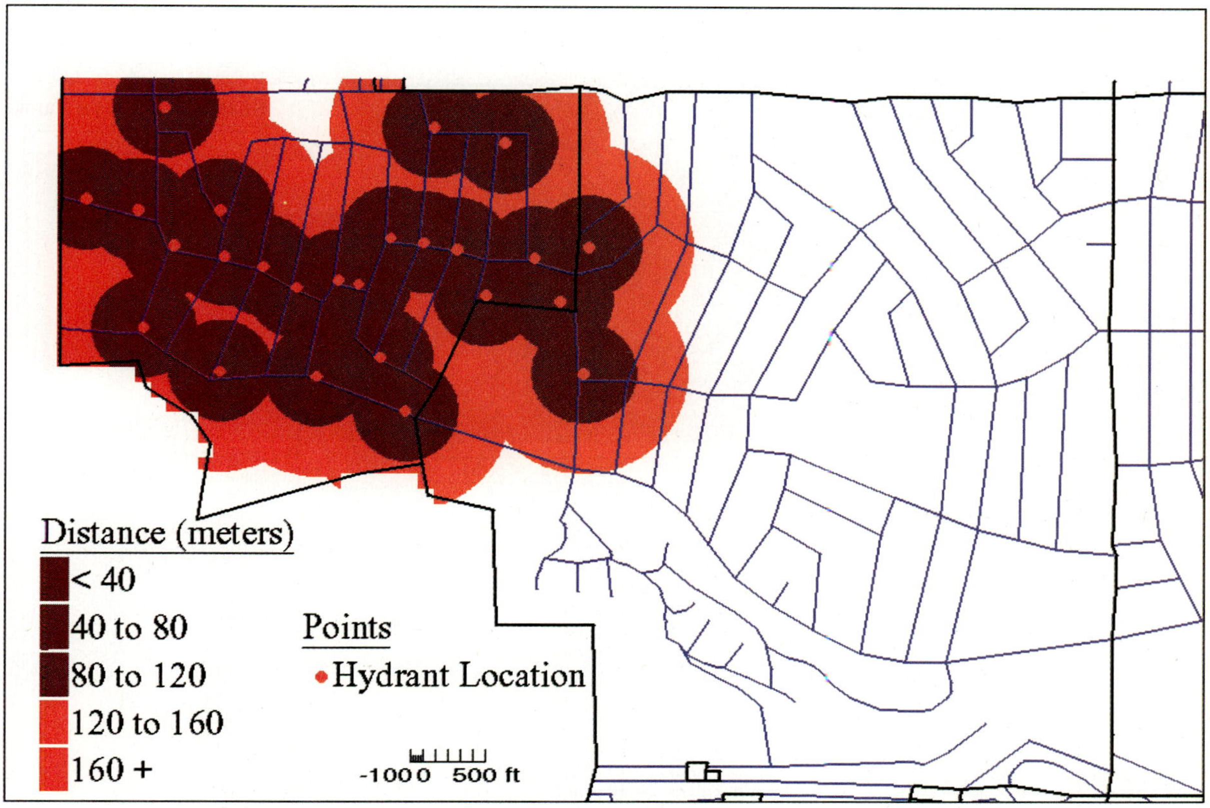

Figure 4.9

Presentation of Maps, Chart, and Spreadsheet

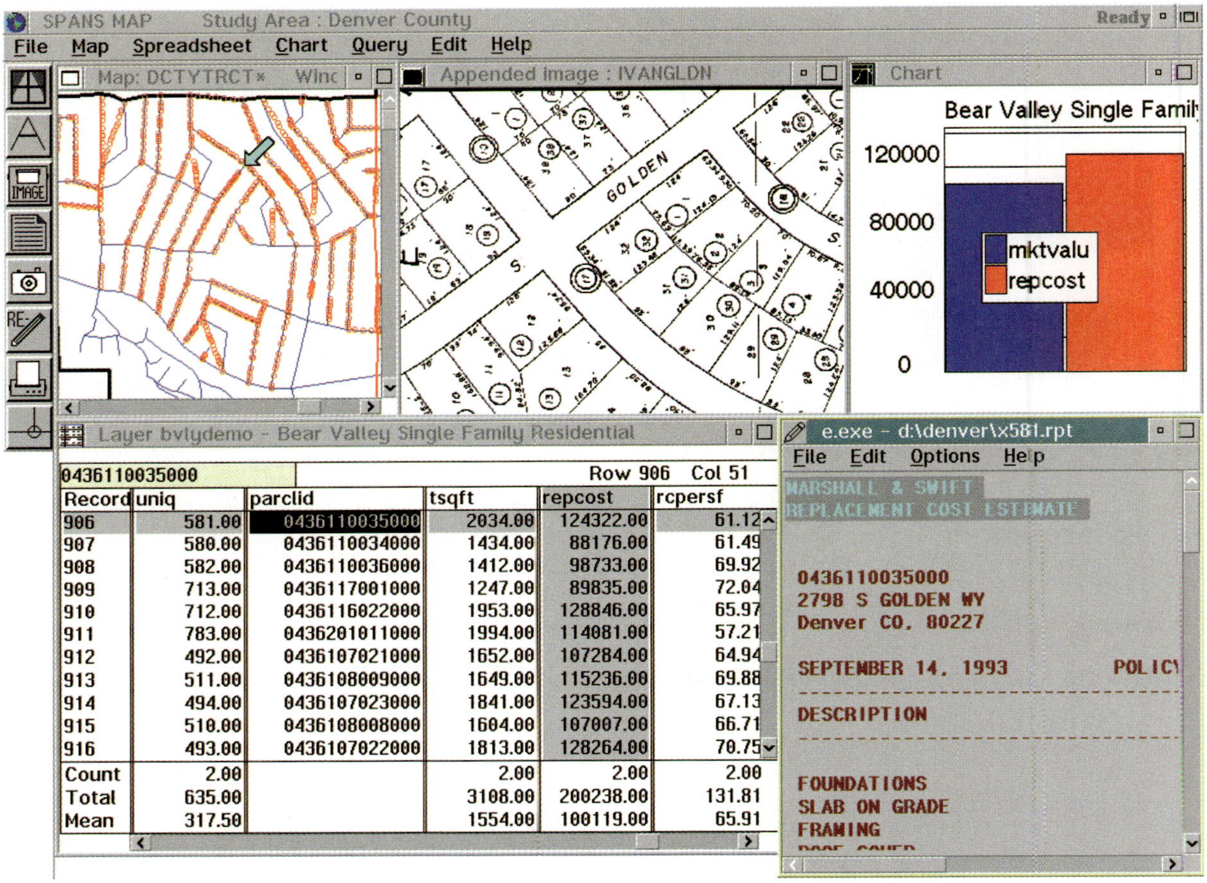

Figure 4.10

Presentation of Cost and Sales Comparison Data

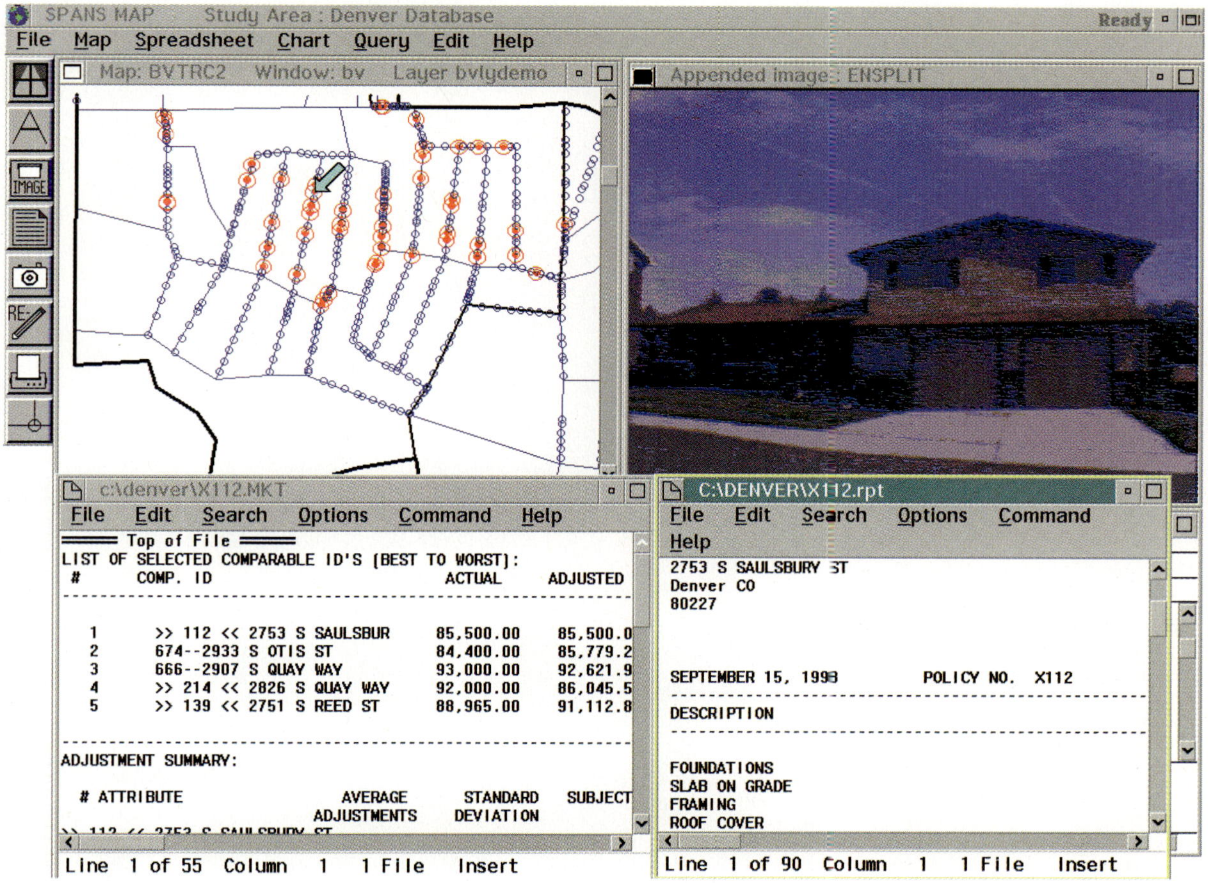

Figure 4.12
Percent of Slope

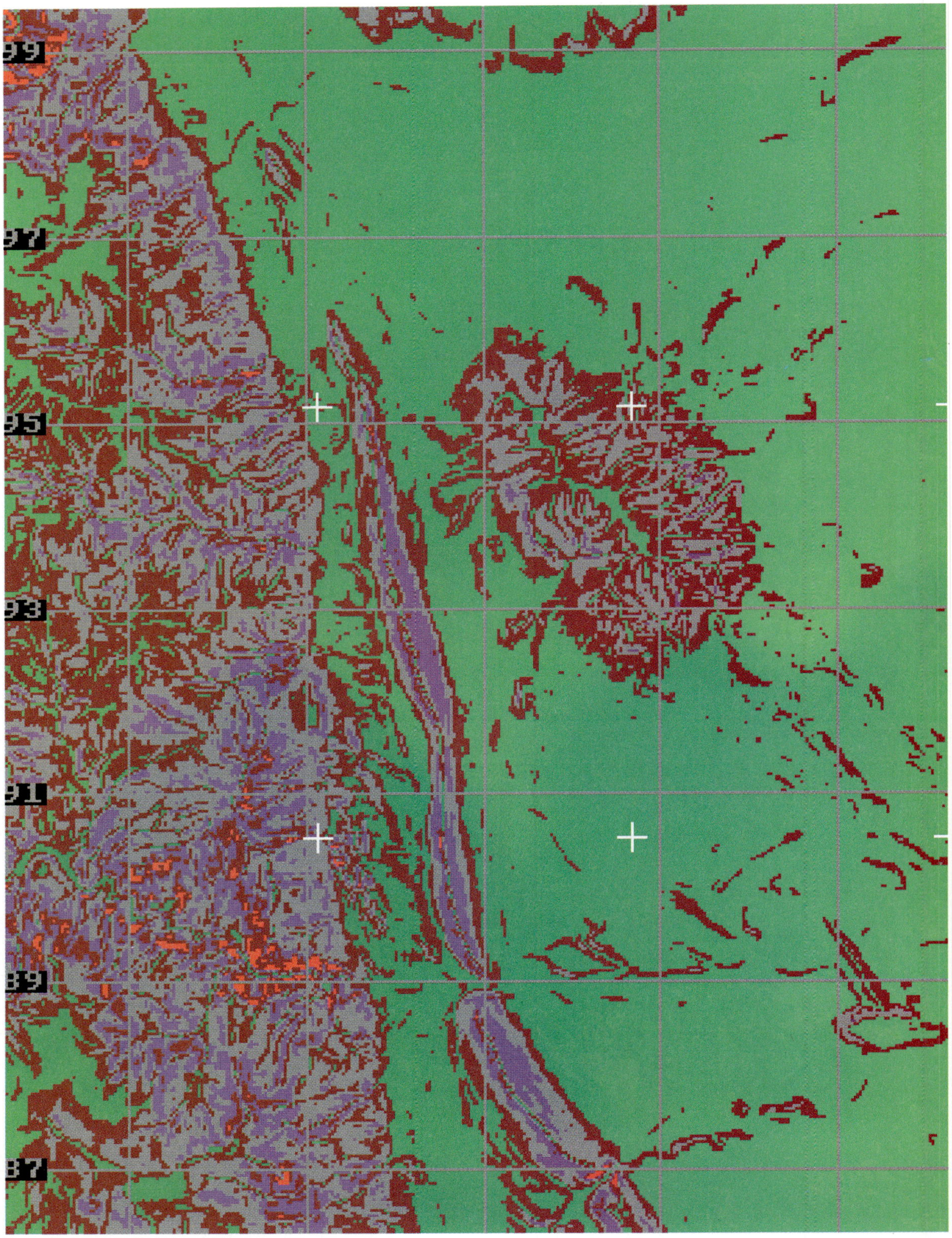

Figure 4.13
Slope Aspect

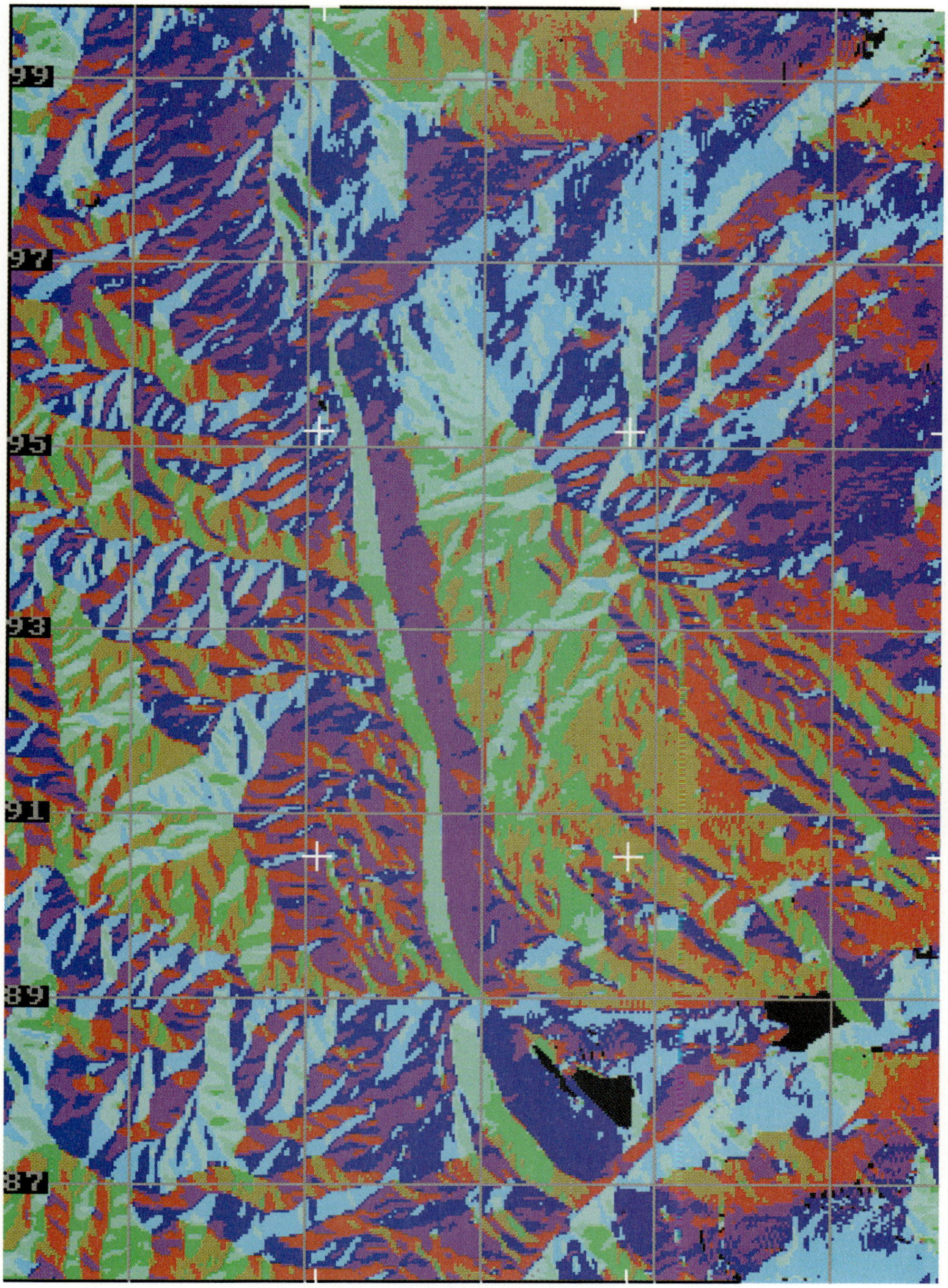

Figure 4.15

Enlarged Portion of Morrison Quad

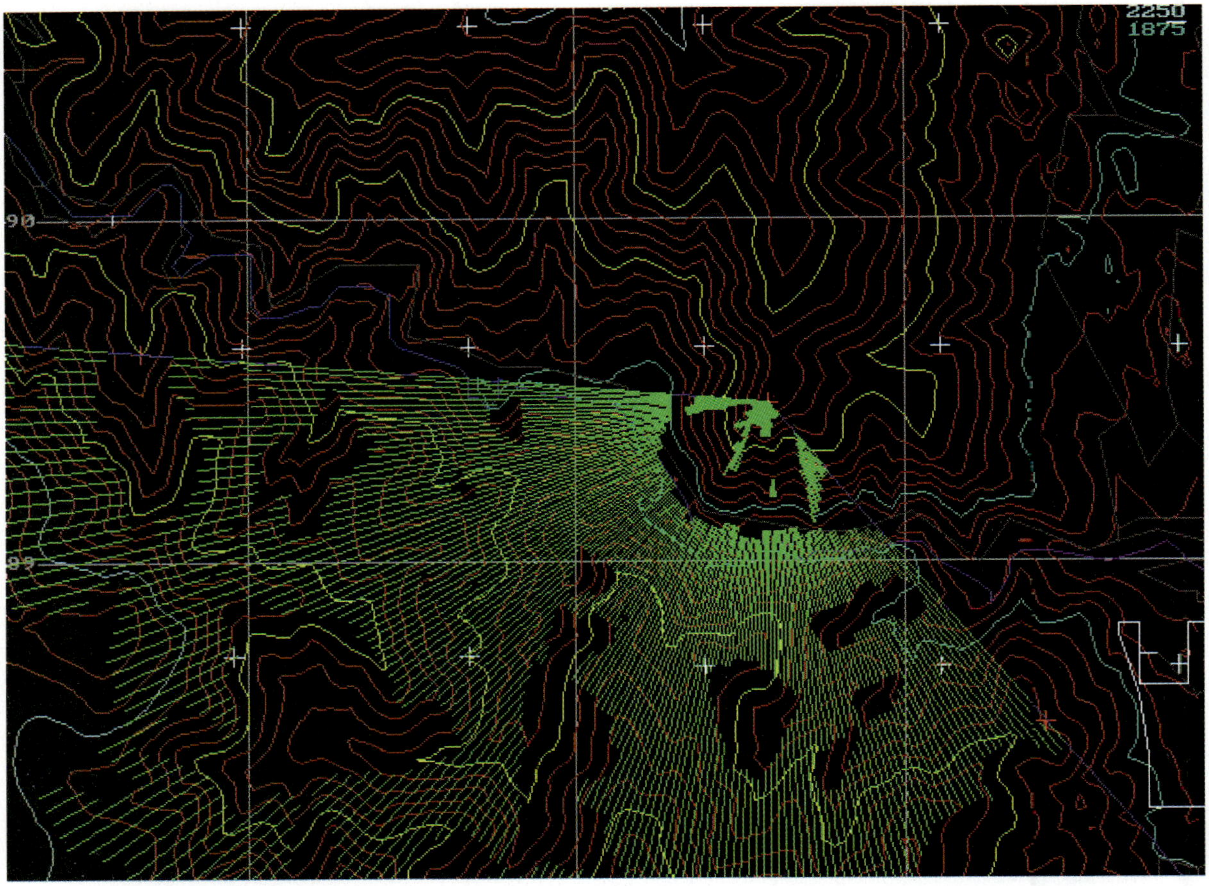

Figure 4.17

Macro Neighborhood Display

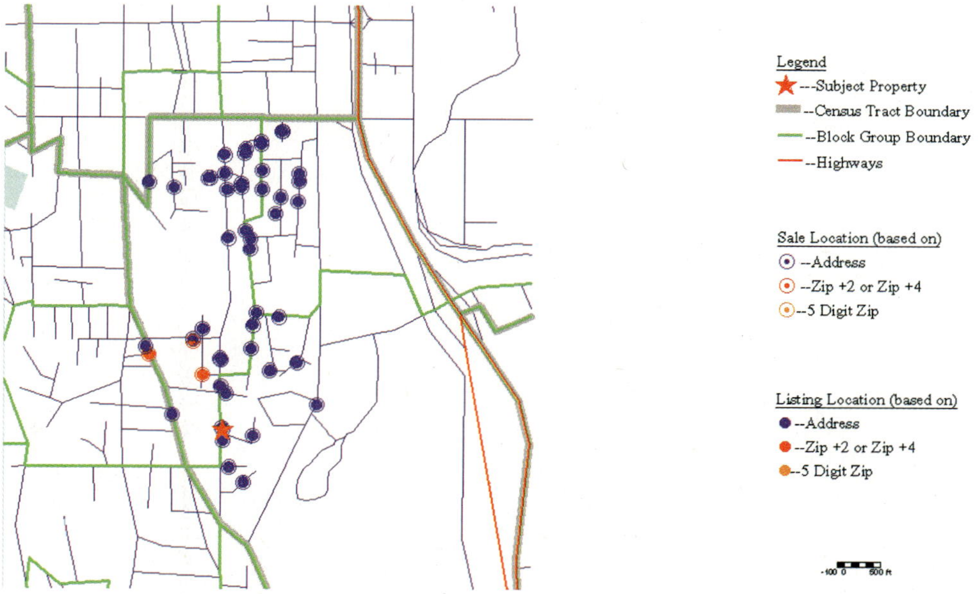

Figure 4.18

Micro Neighborhood Display

Figure 5.5

Overlay Map of Sales by Price per Acre

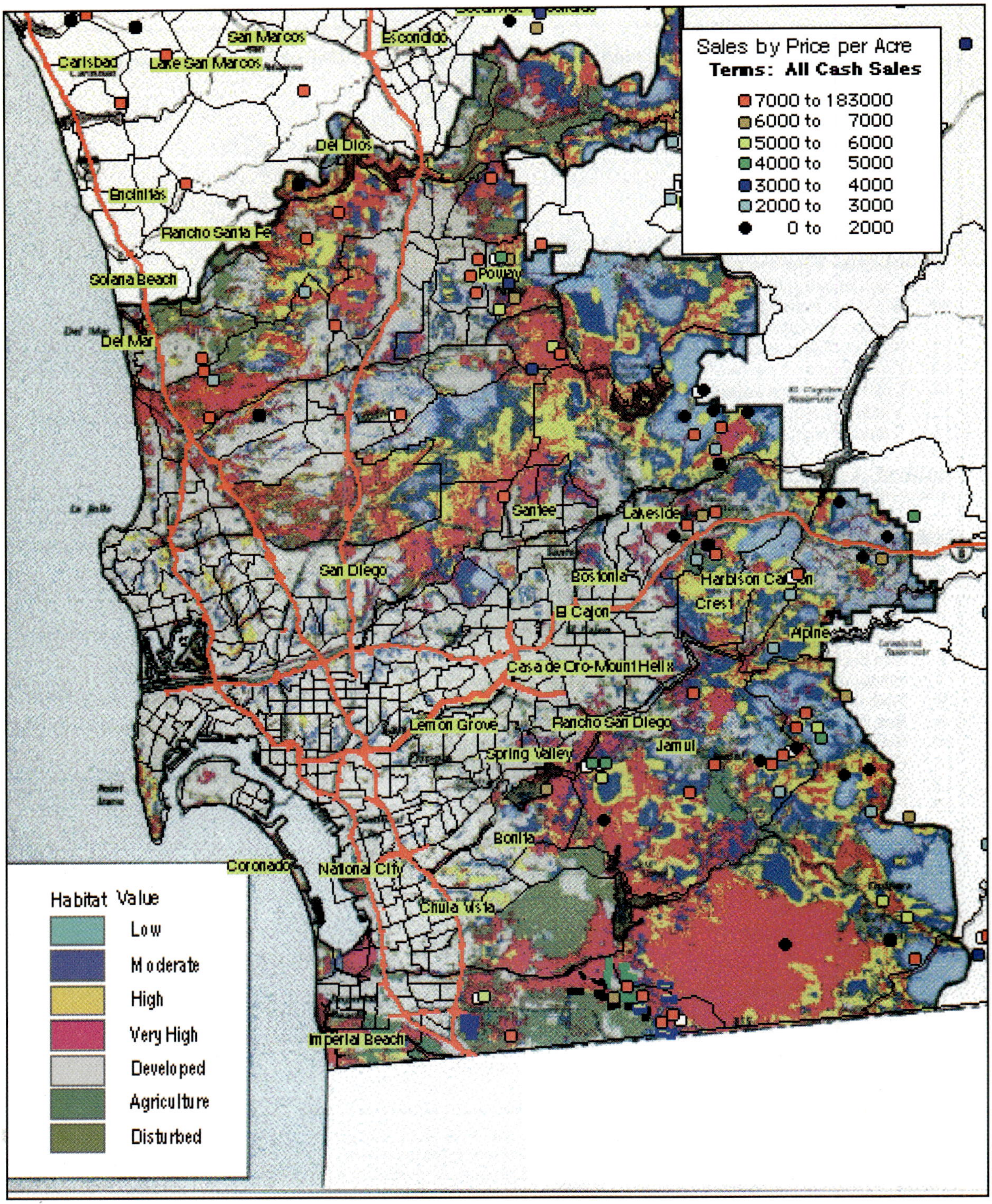

Figure 5.6
Supply and Demand by Biological Submarket

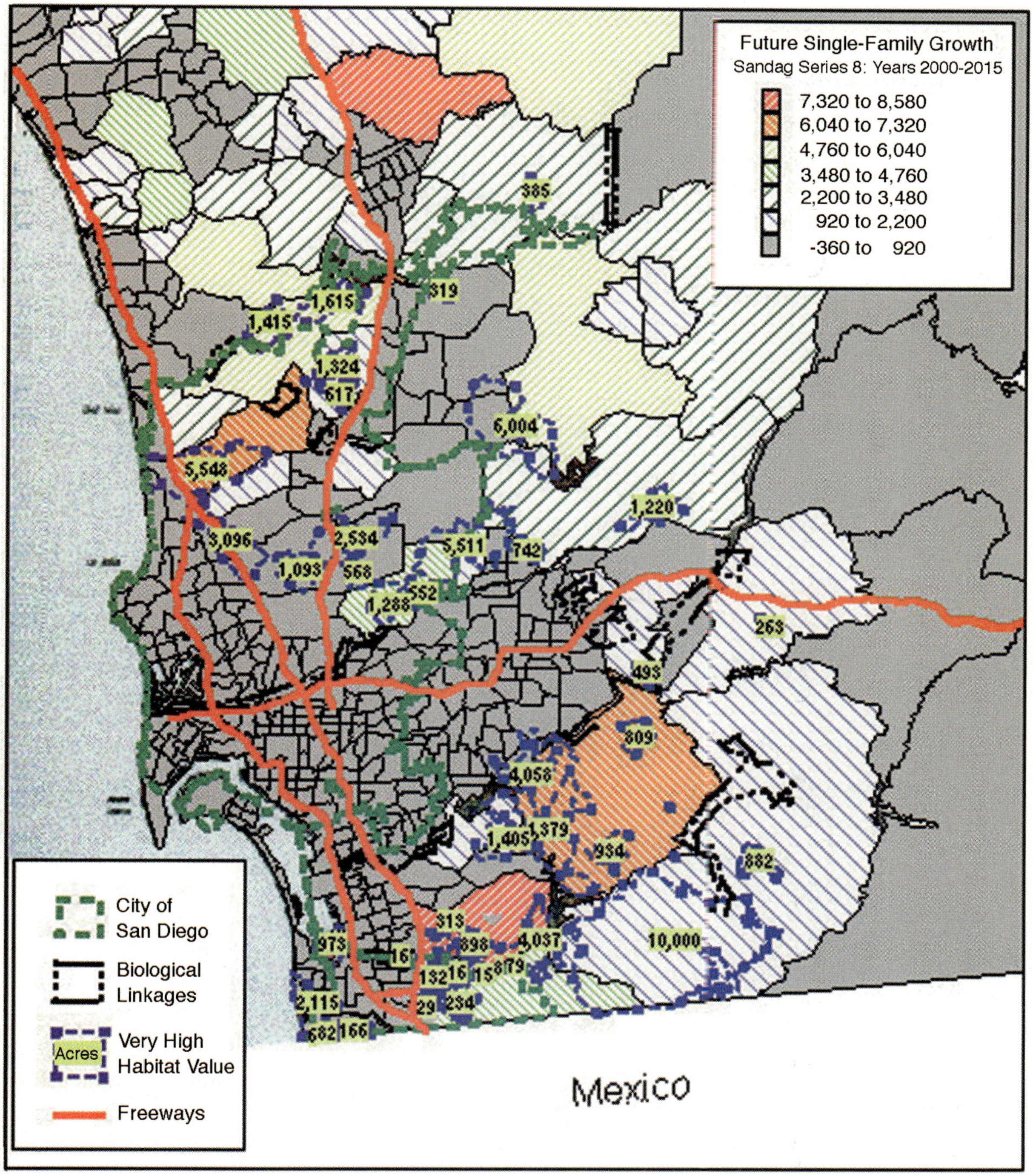

Figure 5.7

Example of Noncontiguous/Nominal Overlays

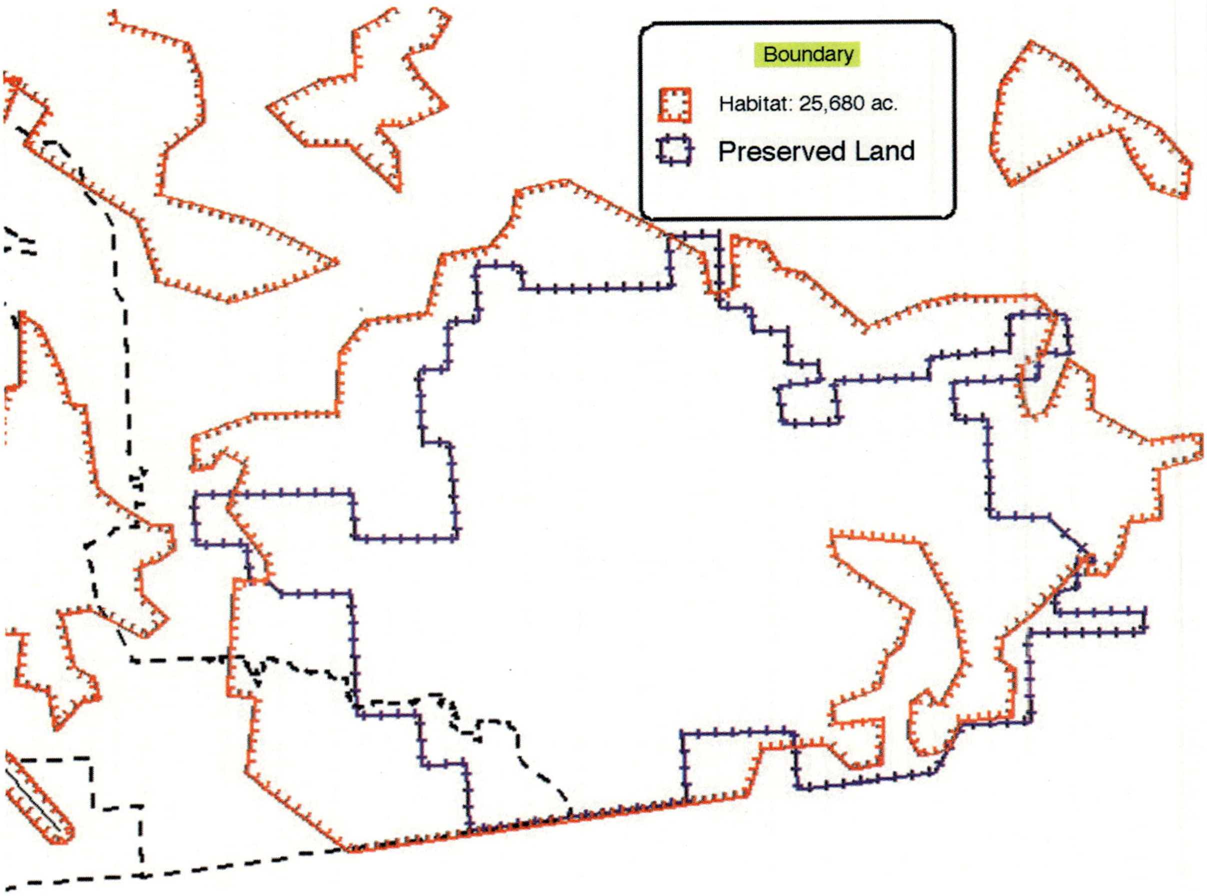

Figure 5.8

Use of BLM "Cutter" to Find Supply of Land

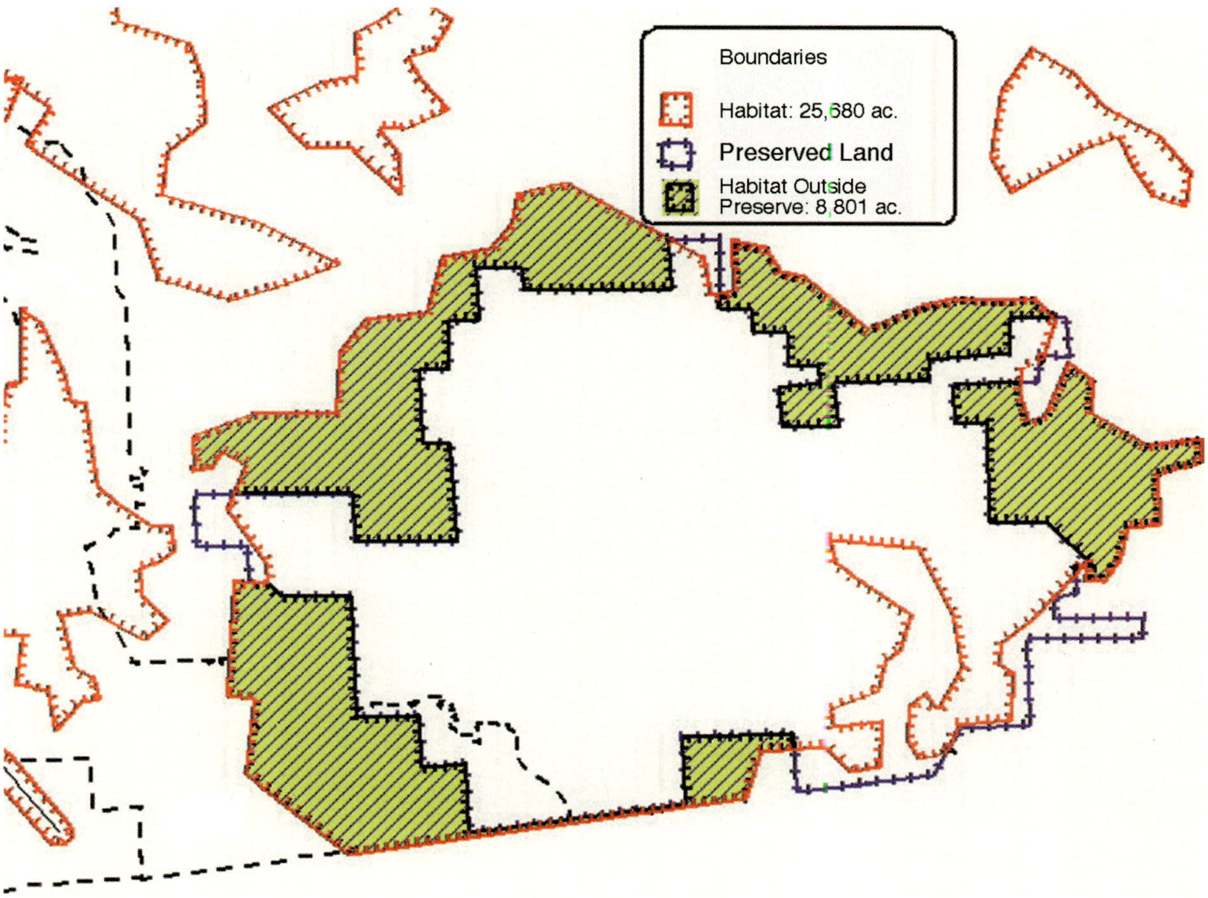

Figure 5.11

Aggregating Sales for Trade Areas A and B

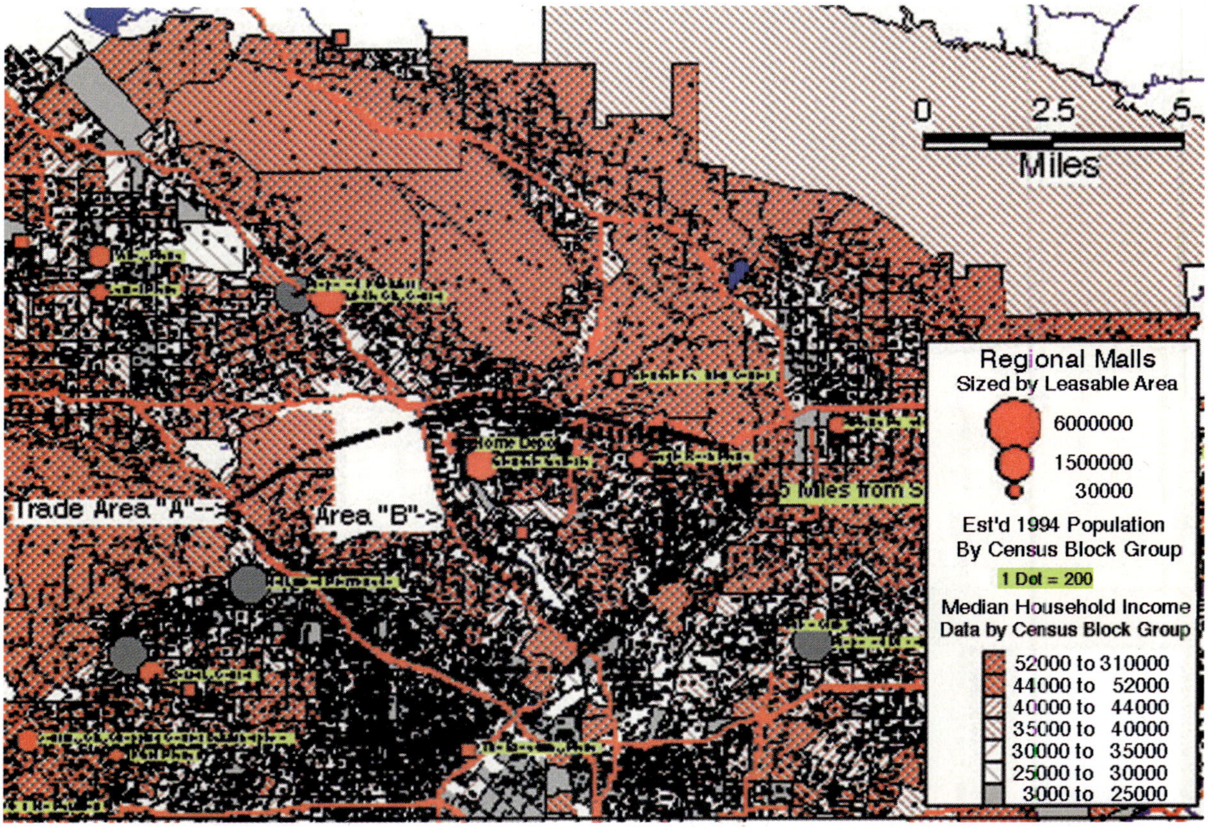

Figure 5.12

"Comparables" with Major Drive Time Population Differences

Site C has twice as many people within a 3 minute drive time as Site B and three times as many people as Site A

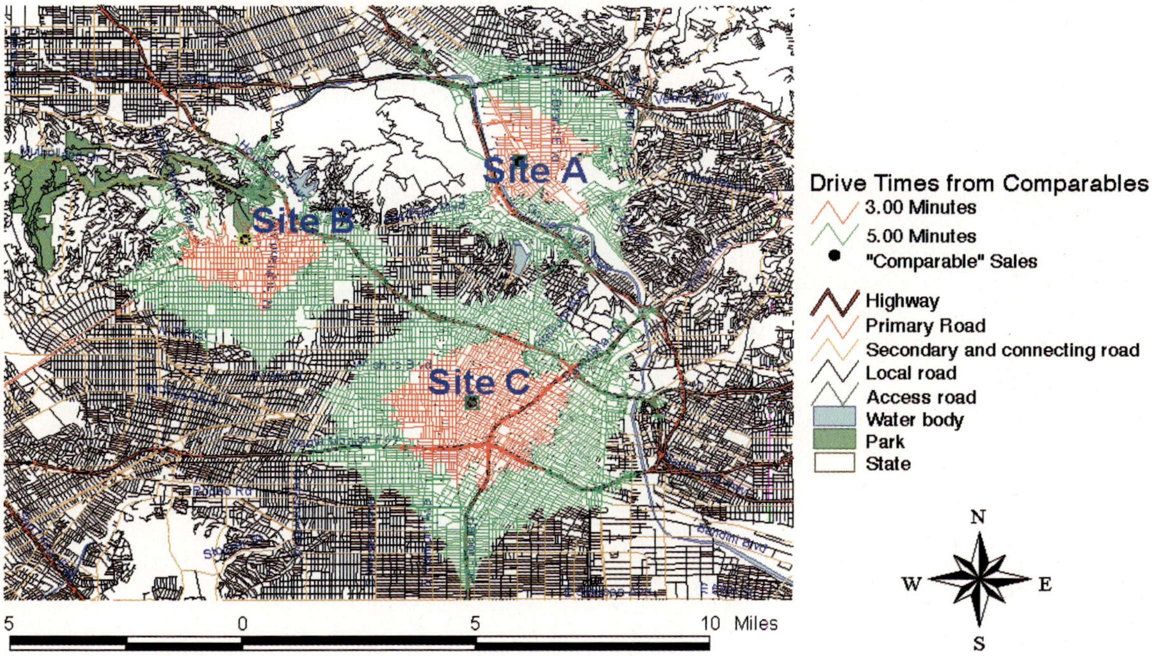

Figure 5.13

Using Census Data and GIS for Exploratory Analysis

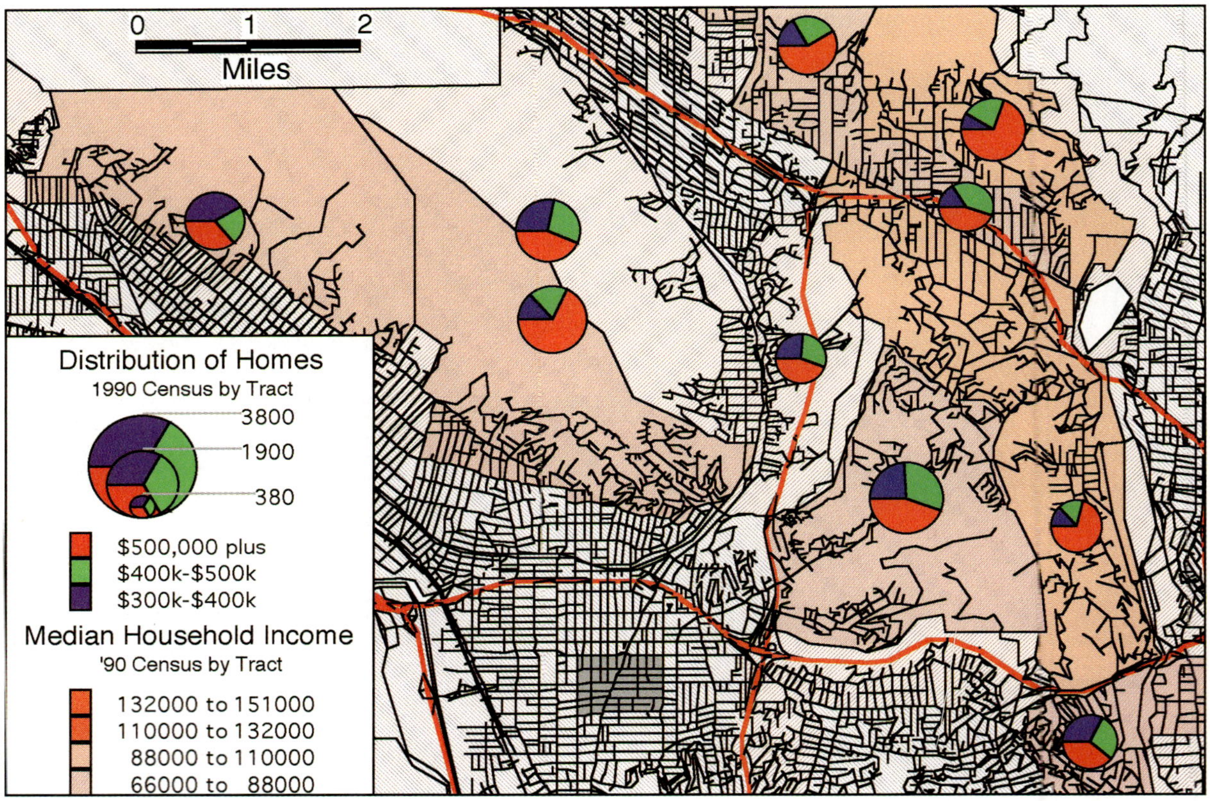

Figure 5.14

Updated Demographics Differentiate High Incomes

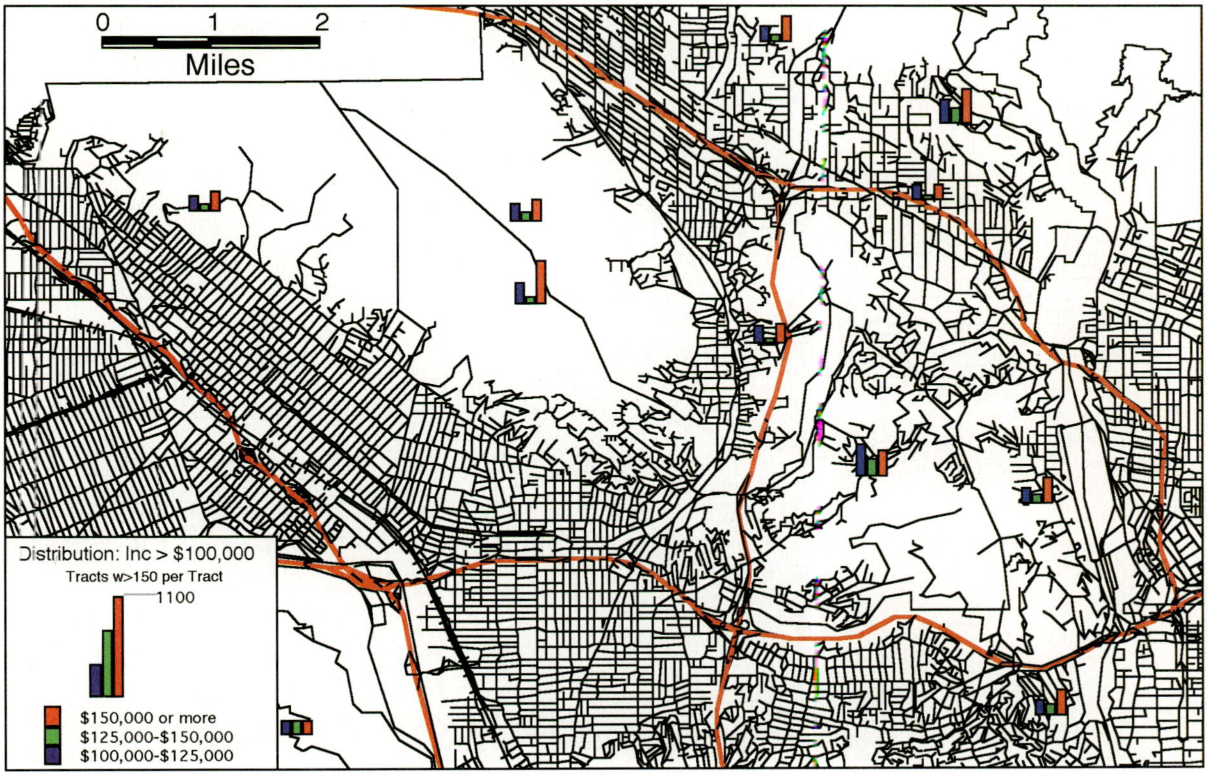

Figure 5.15

An Income Database to Locate Competitive Areas

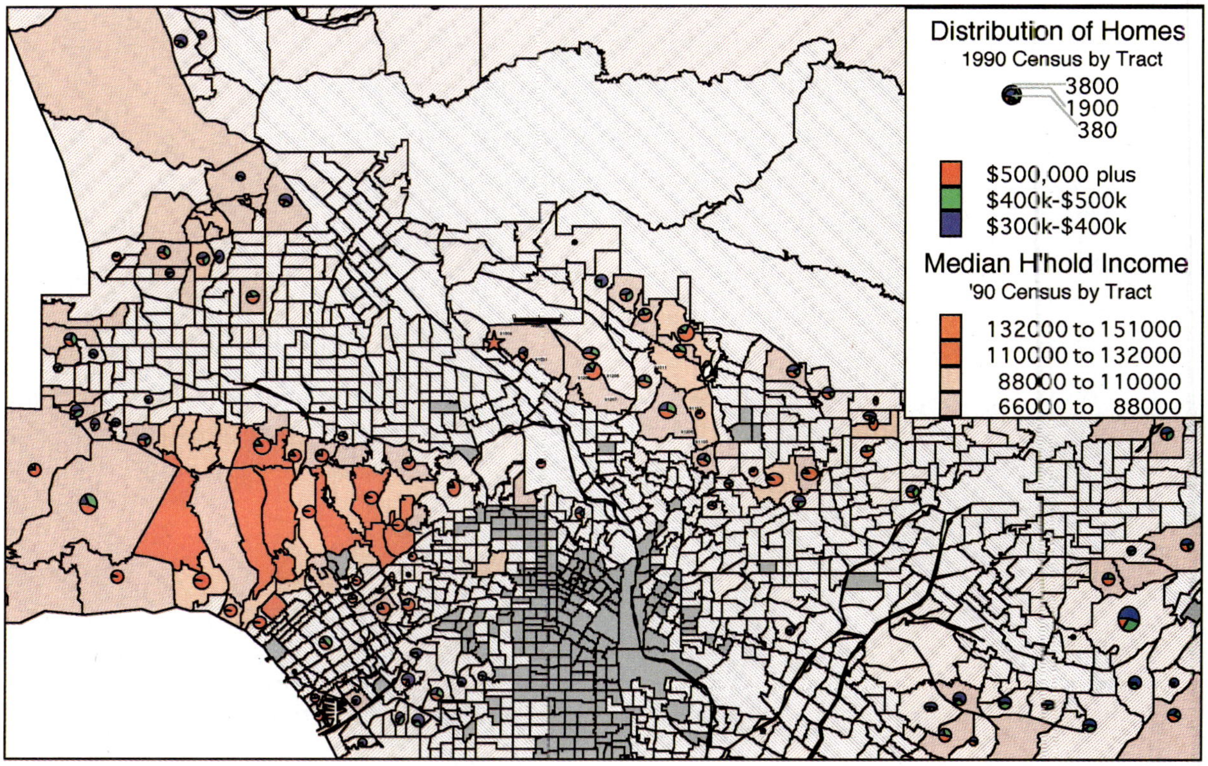

Figure 5.18

Streets Within a 34-Minute Drive Time of Subject Site

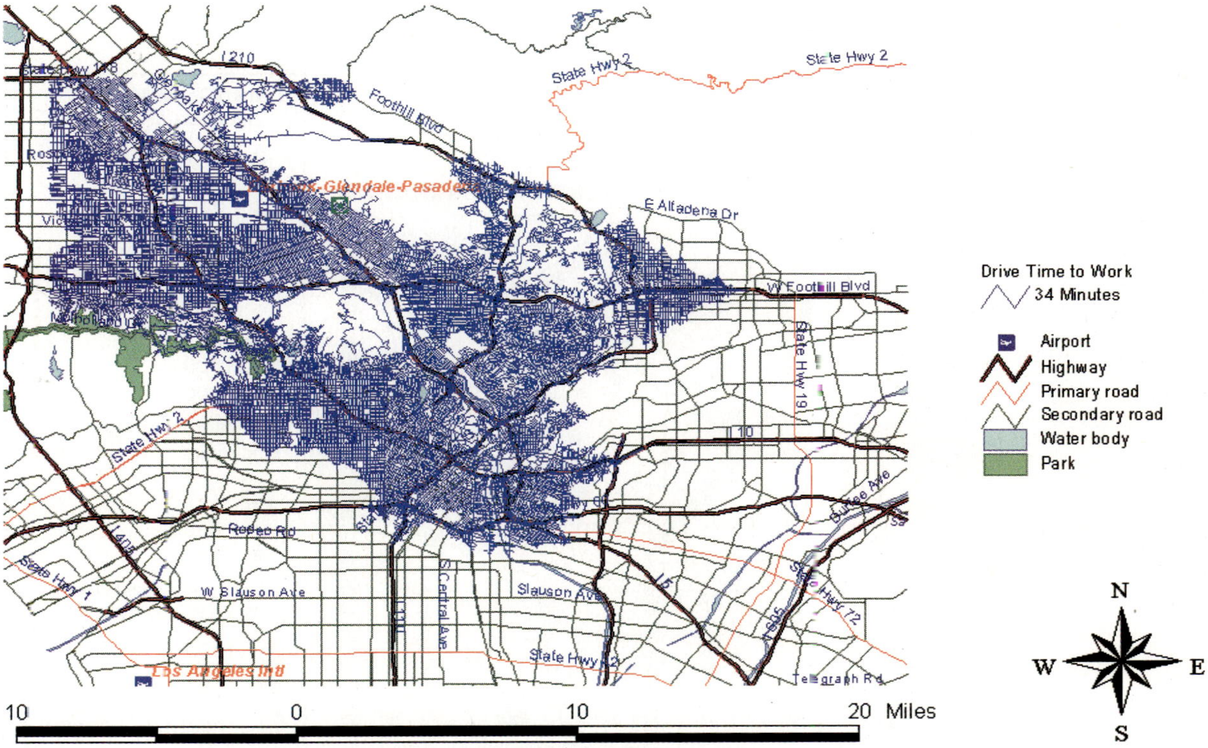

Figure 5.19

Using Disparate Databases to Define a Market Area

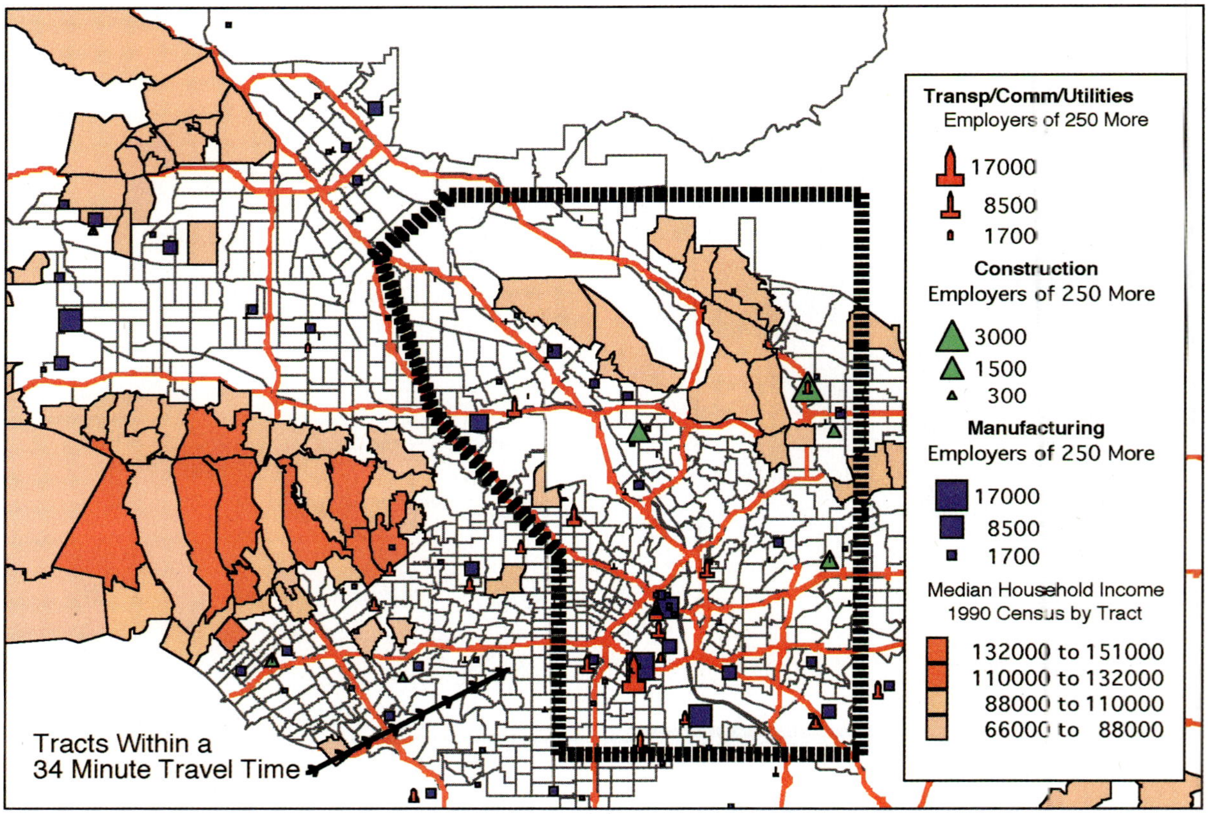

Figure 5.21

Delineation of Primary and Secondary Market Areas

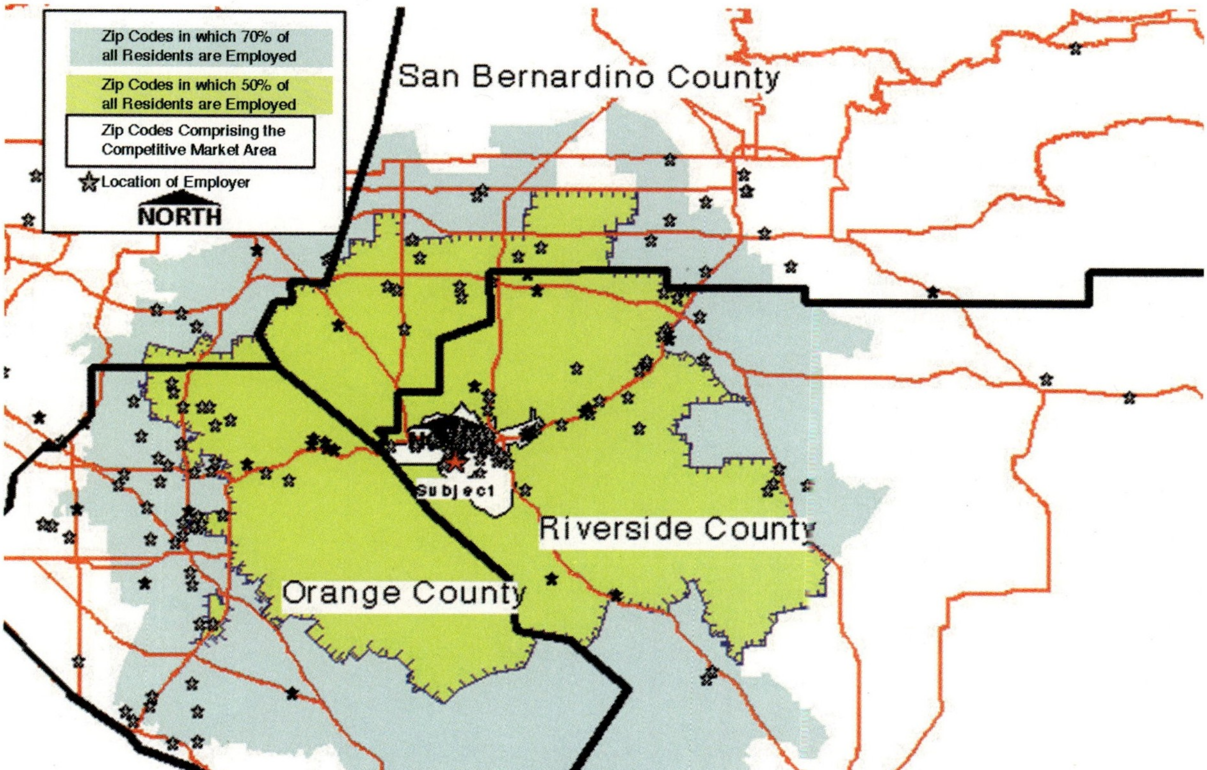

Figure 5.22

Mapping Changes in Demand

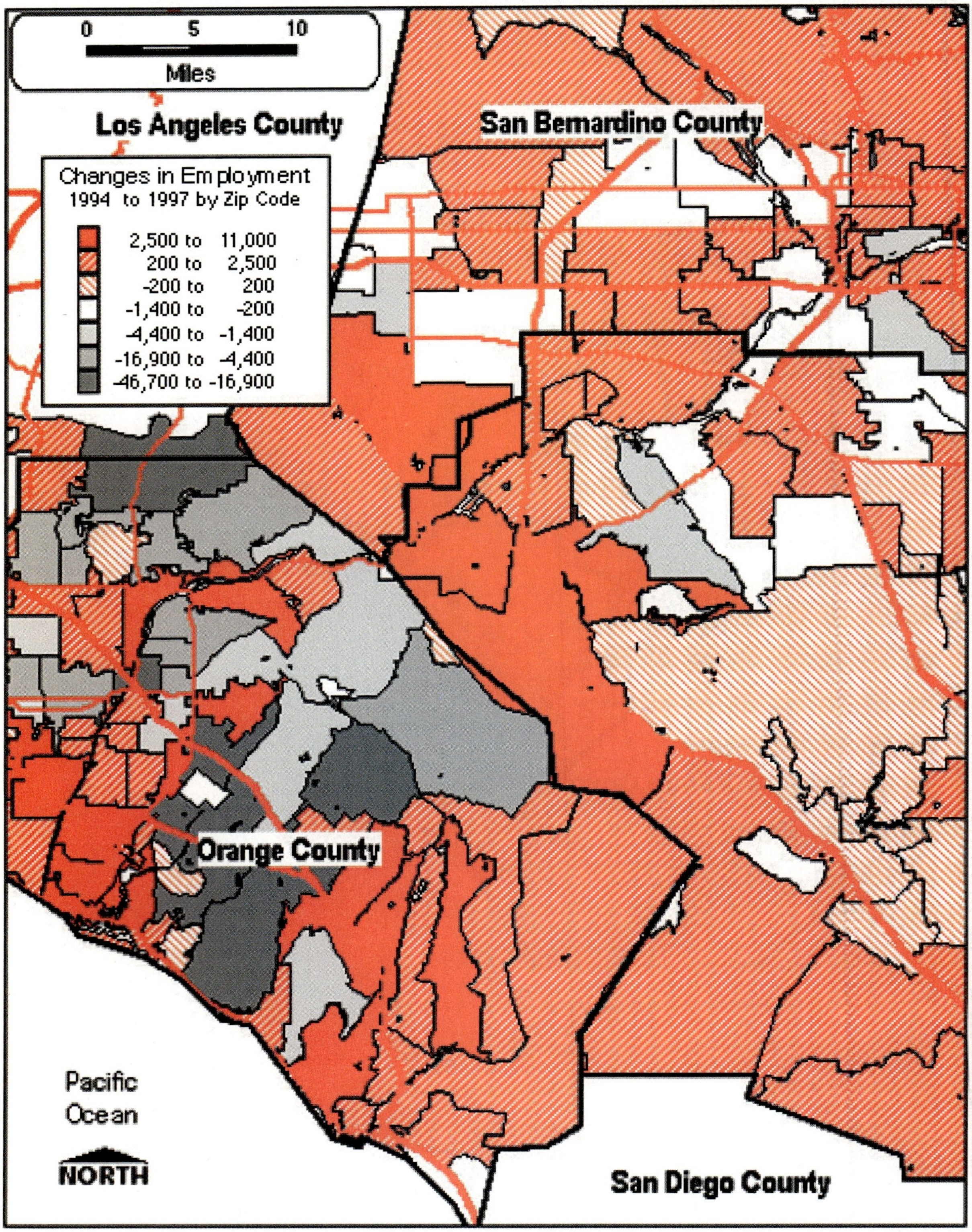

Figure 6.1

Simple Software-Produced Map

Figure 6.2
Neighborhood Map

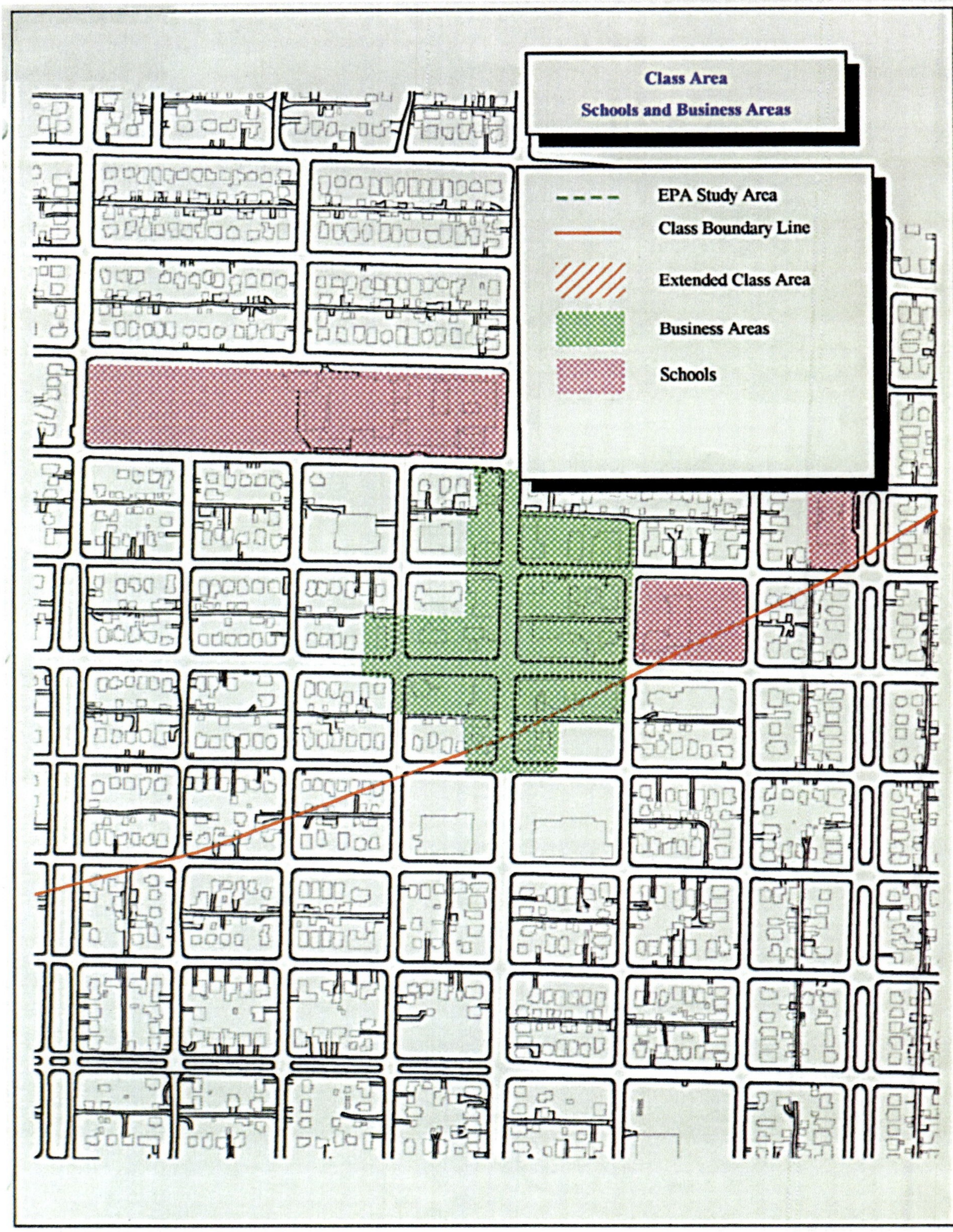